SOUND MATERIALS

A Compendium of Sound Absorbing Materials for Architecture and Design

TYLER ADAMS

FRAME

CONTENTS

INTRODUCTION

When sound collides with a surface, it interacts with and responds to the material and formal conditions of that surface. These interactions, however subtle they may be, are conveyed to the listener when sound reflects from the boundaries of our environment and arrives at our ears. All materials reflect or absorb sound in their own particular manner.

Materials designed specifically to absorb sound may be found in a wide variety of everyday environments — classrooms, subway stations, lobbies, airports, museums, courtrooms, auditoriums, offices, factories, and swimming pools. Sound absorbing materials are used to reduce noise, to improve speech communication, to enhance the listening experience of music, and control reverberation. Often the acoustic design and engineering of these environments goes unnoticed. It is only when sound absorbing materials are absent, when spaces become raucous and speech becomes unintelligible, that our aural experience is foregrounded.

Sound absorbing materials are found in many everyday objects — dishwashers, power tools, washing machines, air conditioning units, automobiles, airplanes, and boats. In these applications, materials can be used to make these devices quieter or to alter their sound quality. The field of product sound quality studies how the sound of products informs customer satisfaction or conveys information that may influence the value of their product. One area where this is extensively applied is automobile design. Studies have shown that even the sound of a car door closing, one of the first things you hear when getting into a car, will establish the first impression of the level of quality, design, and power. As a result, some manufacturers have focused significant efforts on designing this basic element.

Sound absorbing materials must be designed and engineered to suit the needs of different applications and environments. A material that is used outdoors, for example to quiet noise from a highway, must be designed to withstand sun, wind, rain, and resist rot and infestation. A material used in a commercial kitchen would likely need to be non-toxic, light-reflecting, and cleanable.

Sound absorbing materials are incredibly dynamic in the sense that the same piece of material can perform differently just by changing the way in which it is installed or mounted. This variability opens up a number of possibilities for the same material to be used in different contexts to solve a variety of problems.

Sound absorbing materials also present a variety of aesthetic possibilities to the designer. They may be installed as a colourful, prominent, visual design element or may be designed to seamlessly blend in with the surroundings. In terms of aural aesthetics, sound absorbing materials can be used to define our impression of spaces, and they may help establish or emphasise spatial contrasts — lively to deadened, public to private, cool to warm.

A plethora of materials can be transformed into sound absorbers by the simple act of perforation — woods, plastics, metals, and stones. While perforation has been understood for many decades, there continues to be room for innovation and digital fabrication technologies are making the process of customization, prototyping, and fabrication easier, more cost-effective, streamlined, and elaborate.

Digital modelling capabilities now allow for complex material designs to be parametrically modelled and their acoustic performance calculated in real-time. Materials can be placed within virtual environments that may be auralised to allow one to hear the acoustic environment and make changes to materials and their placement within the environment accordingly. Following the iterative process of modelling and auralisation, material data can be fed directly to a CNC machine for immediate fabrication. A number of intersecting technological and cultural shifts present a rather exciting time for sound absorbing materials. On the materials science front, new materials continue to emerge such as wood and metal foams, aerogels, and activated carbon. New methods of weaving and textile processing have resulted in the translucent acoustical textiles. Issues of sustainability are prompting many to explore the use of new kinds of plant and animal fibres or simply reformulate the manufacturing processes of existing materials.

While there are many publications that catalogue various kinds of architectural and design materials, to date there has never been such a publication dedicated to sound absorbing materials. This lack of information may contribute to a common perception that fabric wrapped fibreglass and lay-in tile ceilings are the only options available. This book aims to show not only the variety of materials currently available but also demonstrate the infinite design possibilities, particularly when collaborations take place at the intersection of acoustic engineering, materials science, design, and architecture.

This book is intended as a digest, providing architects, designers, and creative professionals with a survey of materials past, present, and emerging, while introducing the very fundamental technical parameters that concern the performance of sound absorbing materials. This book does not intend to supplant the expertise of a seasoned acoustician. To the contrary, it is hoped that this book will encourage future collaborations and conversations between designers and engineers and in a small way advocate for greater consideration of, and enthusiasm for, acoustics in the design of everyday objects and environments.

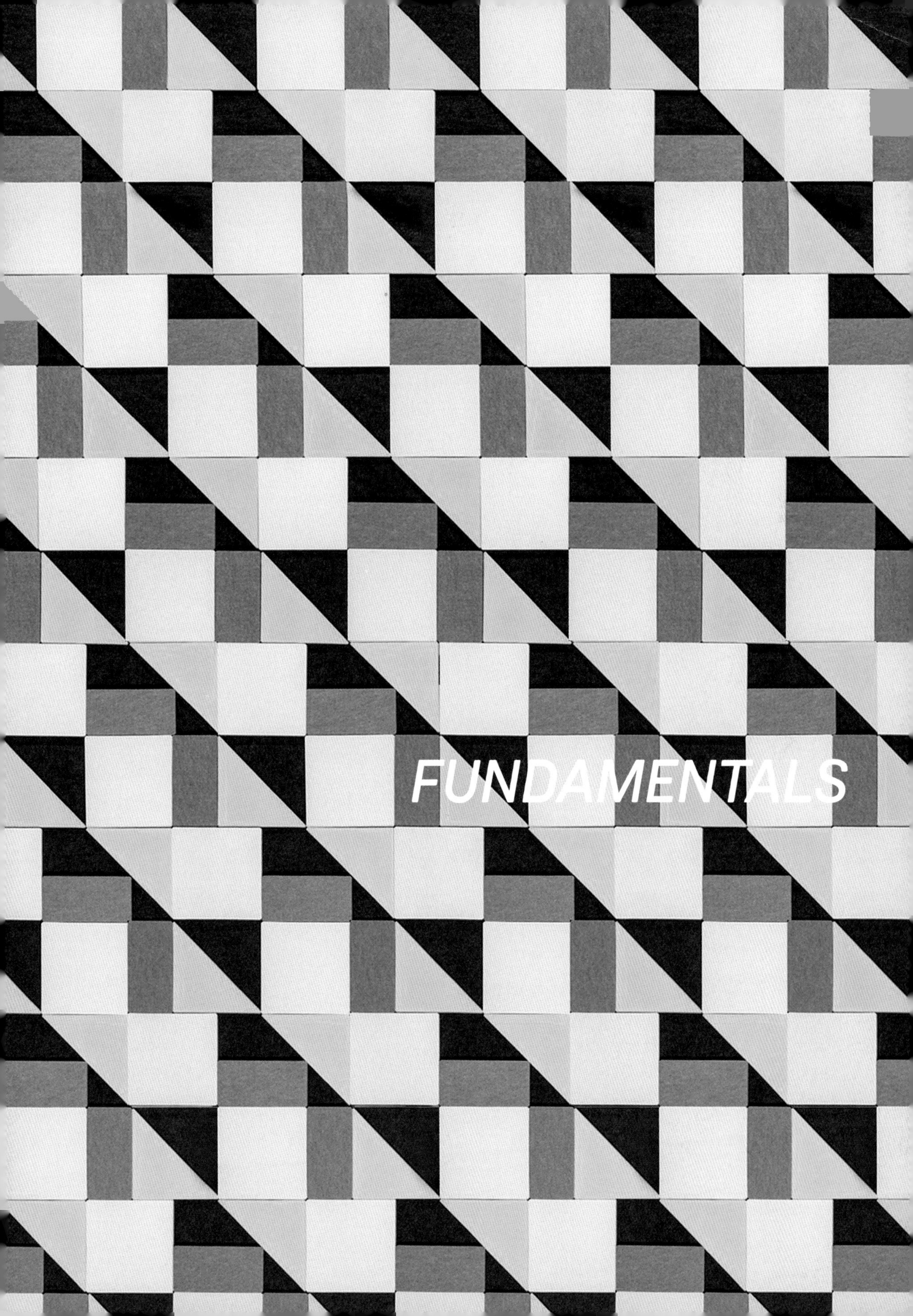
FUNDAMENTALS

OVERVIEW

Acoustics is a vast and complex field of study which encompasses a variety of disciplines such as speech communication, psychology, mechanical engineering, physics, oceanography, medicine, physiology, architecture, electronics, music, and the arts.

From a very fundamental level, the acoustic environment can be broken down into three elements: the source, the path, and the receiver. The source is the thing which is producing sound — a human, a bird, a speaker, etc. The path is the medium through which the sound transmits — this may be the air or it could be the structure of a building. The receiver may be a listener or a microphone. If a sound source is considered undesirable, then the design objective is to reduce the ability for that sound to reach the receiver. This may be accomplished by creating distance between source and receiver or manipulating the environment or path to attenuate or block the sound. If a sound source is considered desirable, such as music or speech, then the design objective is to maintain clarity or enhance the sound source so that it is clearly received and not distorted or degraded by the environment.

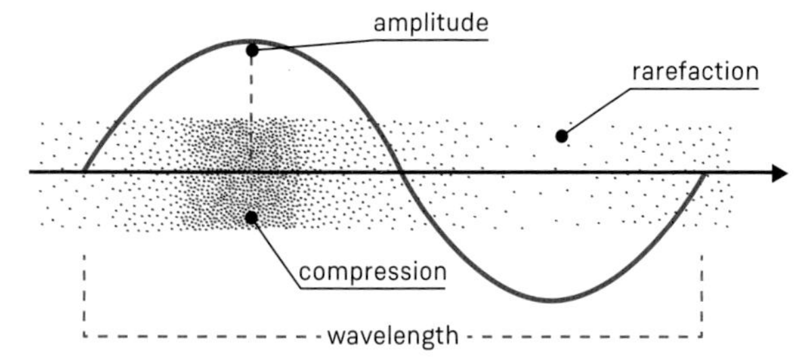

Sound is a mechanical wave that propagates through air by way of compression and rarefaction of molecules. The rate of change constitutes frequency [Hz]. The amount of change in pressure constitutes amplitude [dB].

In physical terms, a sound is a mechanical wave that propagates through some medium (solid, liquid, gas) by way of compression and rarefaction of molecules in that medium. In the air, this results in microscopic changes in the local atmospheric pressure. The rate of the changes in pressure in one second constitutes the frequency, expressed in Hertz (Hz). The amount of change in pressure constitutes the amplitude, expressed using the logarithmic decibel (dB) scale.

All sounds possess these fundamental properties of frequency (what we may perceive as pitch) and amplitude (what we may perceive as loudness).

Frequency can be described in terms of wavelength, the spatial distance over which the wave's shape repeats itself. The lowest frequency we can hear (20 Hz) has a wavelength of roughly 17 metres, whereas the highest frequency we can hear (20,000 Hz) has a wavelength of 0.02 metres. It is important to understand wavelength because the degree to which an object or surface interacts with sound is dependent upon its dimensions in relation to wavelength. As is shown in the porous materials overview, absorption performance is directly related to the thickness of material in relation to wavelength. Wavelength can be calculated using the following simple equation:

$$\text{wavelength } [\lambda] = \frac{\text{speed of sound [c]}}{\text{frequency [Hz]}}$$

The speed of sound varies dramatically according to the elasticity and density of the medium. In air, the speed of sound can change according to temperature and humidity but is generally considered to be around 343 metres per second (or 1125 feet per second).

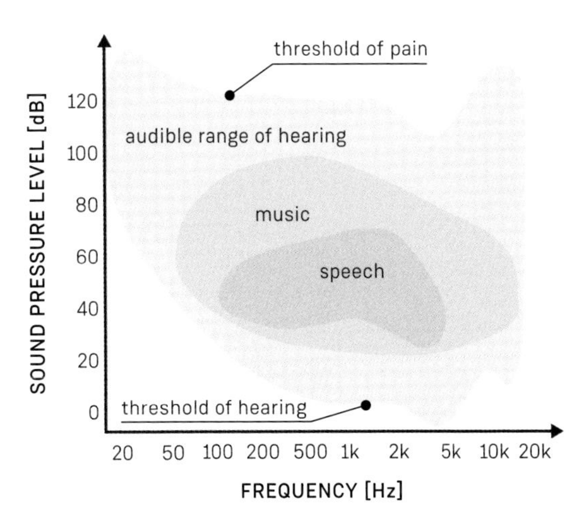

The audible range of hearing is shown overlaid with the typical ranges for music and human speech sounds. The ear does not hear all frequencies equally and is less sensitive to very low frequencies, below 100 Hz.

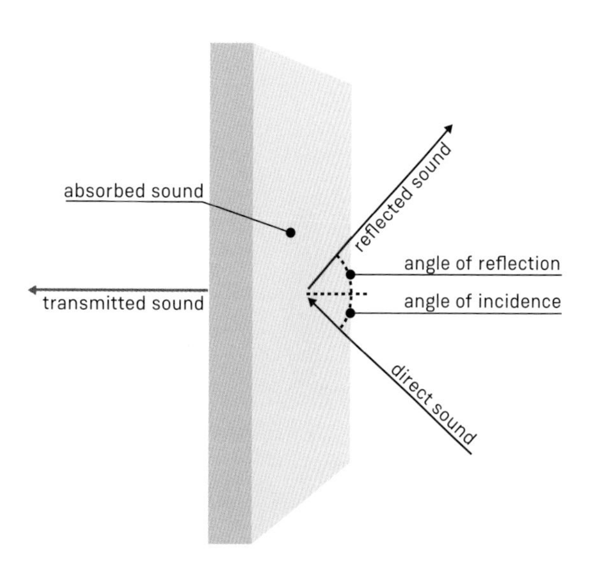

When sound strikes a material, a fraction of energy is absorbed, reflected, and transmitted through. For flat surfaces, the angle of reflection will equal the angle of incidence.

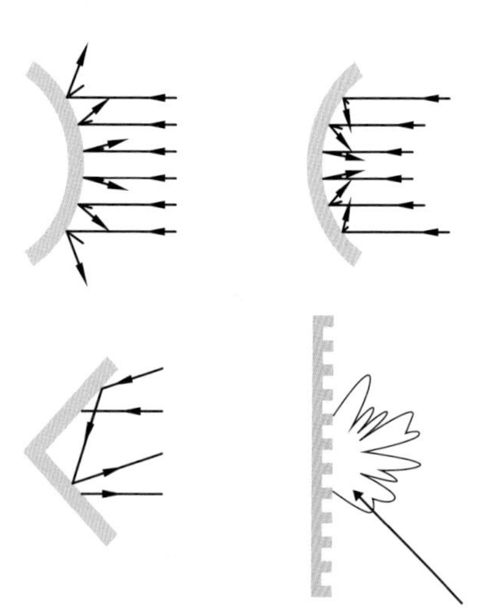

Diagrams showing the different ways in which the form of a surface can cause influence sound reflection. Convex and ornate surfaces will cause sound to scatter, flat surfaces cause specular reflections, and concave surfaces cause focusing.

The sounds we hear are typically a complex mixture of many frequencies at different amplitudes which are varying over time. When taking acoustical measurements or conducting calculations, for the purpose of simplification, it is common to divide the frequency spectrum into 1/3 octave bands or 1/1 octave bands. Acoustic materials are commonly measured in 1/3 octave band resolution and then simplified into 1/1 octave bands or single number metrics.

There are three fundamental interactions that take place when a sound impinges upon a material: some energy is reflected back, some energy is absorbed into the material (converted into a small amount of heat), and some energy may be transmitted through the material to the other side. All materials absorb, reflect, and transmit sound in their own particular way. Just as the material composition of an environment establishes visual identity, it is may also form an aural identity. Sound absorbing materials are typically used to manipulate the environment, either to quiet a source of noise or to enhance the reception of sound.

In concert with material, sound is also influenced by form. At the scale of surface, the complexity of texture may cause a sound to reflect more diffusely, while a flat, monolithic surface will reflect specularly. This interaction is also dependent upon the sound's wavelength in relation to the dimensions of the surface patterning. A diffuse reflection scatters the sound in various directions, avoiding repetitive reflections between parallel surfaces and resulting in a sound field that is more homogeneous. Baroque architecture, with ornate finishes and sculptures, often results in diffuse sound environments.

Smooth surfaces shaped into concave parabolic forms can focus sound into beams, amplifying and transmitting that sound a greater distance. This is often experienced in 'whisper-galleries' where the sound of a whisper seems to cling to the edge of a wall and reappear elsewhere. Smooth convex forms will scatter or diffuse sound reflections.

When surfaces come together to create a sense of enclosure, sound is supported, amplified, and filtered through the phenomenon of resonance. A resonance is a tendency for certain frequencies to be emphasized due to the volume, form, and material conditions of a space. Resonances occur at various spatial scales. When we blow across the top of an empty glass bottle, we excite its resonant frequency. In small rooms, these resonances are often referred to as modes, which occur at very low frequencies. Room modes pose problems in recording studios because the resonances can cause some low frequencies to be unusually loud and others to be unusually quiet.

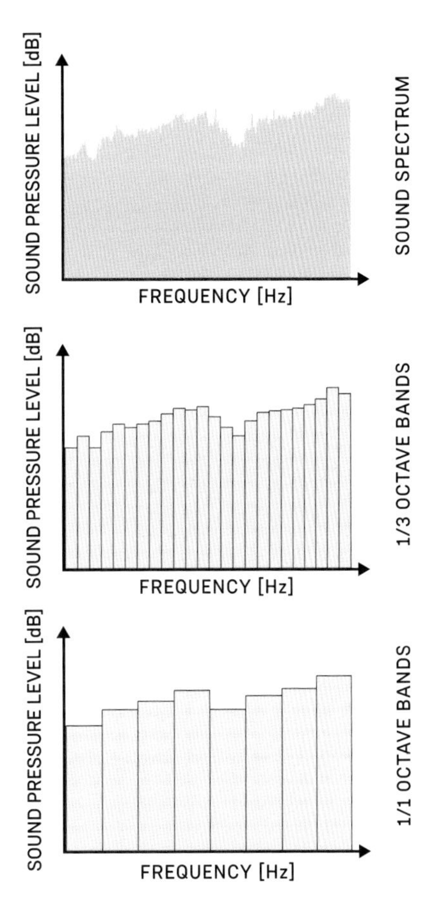

A complex sound spectrum is commonly averaged into 1/3 octave bands or 1/1 octave bands for the purposes of simplification when taking measurements or conducting acoustic calculations.

MATERIAL APPLICATIONS

Sound absorbing materials can be used for a variety of practical applications, namely to reduce noise levels in an enclosed space, to reduce reverberation, to eliminate echoes, and to improve speech intelligibility. These applications can dramatically influence our perception of our environments, which we may describe as clear or muddy; lively or dead; bright or dark; dry or wet; intimate or spacious; rich or thin; coherent or incoherent; harmonious or discordant.

The sounds we perceive consist of both the direct sound from the source as well as the reflections from the objects and surfaces in our environment. When these sounds combine at our ears, they tell us not only about the source but also the environment or context within which the source and receiver are placed. The reflections from the objects and materials in our environment is why our voice sounds different singing in the shower compared to outdoors.

In a free-field where there are no sound reflecting surfaces, only the direct sound is received and sound spreads out spherically from the source. In these conditions, the sound level will decrease by 6 dB for every doubling of distance between the source and receiver.

In most environments, particularly indoors, there are objects and surfaces which reflect sound and these reflections can add to the total sound level received. In a reverberant environment, sound may behave under free-field conditions when very close to the sound source (typically less than 2 metres) but the reverberant field will dominate at greater distances. By adding sound absorption to a room, the reflections from the room surfaces can be reduced or eliminated so that primarily the direct sound is heard.

The reflections that make up the reverberant field will add to the total level of sound. By adding absorptive materials to a reverberant space, you can reduce the reverberant reflections and thereby reduce the overall sound levels within a space. For each doubling of the total amount of absorption in a room, the sound level can be reduced by 3 dB. The following simple formula can be used to predict the reverberant sound level reduction by comparing the total absorption in the room before and after absorbing materials have been added.

$$\text{noise reduction [dB]} = 10 * \log \frac{\text{total room absorption } \textbf{after} \text{ treatment [sabins]}}{\text{total room absorption } \textbf{before} \text{ treatment [sabins]}}$$

The formula is further explained in the Example RT calculation in a following section. For most situations, the upper limit of this noise reduction is 10 dB, which perceptually would sound like a 50 per cent reduction in the total sound level.

In noisy environments, humans tend to involuntarily increase their vocal effort to be heard over the background noise. This is referred to as the Lombard Effect and can be experienced first hand in any loud restaurant or bar. Reducing reverberant noise can help avoid situations where the Lombard Effect occurs.

Reducing sound reflections is desirable for a number of other reasons: reflections can make speech unintelligible, they can make it difficult to locate where sounds are coming from, and they can make music sound muddy. On the other hand, if there aren't enough reflections, a space can feel too dry, quiet or dead, so a delicate balance must be achieved.

The intelligibility of speech is dependent upon a number of factors that include the distance between the source and receiver, the loudness of the source compared to the background noise level, and the amount of reverberation in the environment. Reverberation impacts intelligibility in two ways, it raises the overall background noise level and the late arriving reflections cause language to become distorted or blurred. Sound absorbing materials are used in all manner of spaces where speech intelligibility is crucial: classrooms, courtrooms, lecture halls, and public spaces where announcements from emergency and public address systems must be clearly understood.

Sound absorbing materials are also used to address very specific types of sound reflections which are often problematic, such as focusing that occurs from a concave surface or echoes, which are reflections delayed in time and perceived as separate from the original sound source. Echoes can occur from flat surfaces located 15 metres or more from the sound source. Most auditoriums will have the rear walls behind the audience covered in absorbing materials to prevent echoes from bouncing back to the stage. If absorptive materials are not used, then diffusive materials are used to scatter the sound. Flutter echoes occur when there are two parallel reflective surfaces. Sound bounces back and forth in quick succession, resulting in a zinging or buzzing sound. This type of echo is prevalent in conference rooms, which may have glass walls, whiteboards, large tables, and drywall surfaces in parallel with one another. Flutter echo can be eliminated by adding absorption to one of the parallel surfaces, by angling the surface to eliminate the parallel condition, or by having an ornate surface that will diffuse the sound so that it does not reflect specularly.

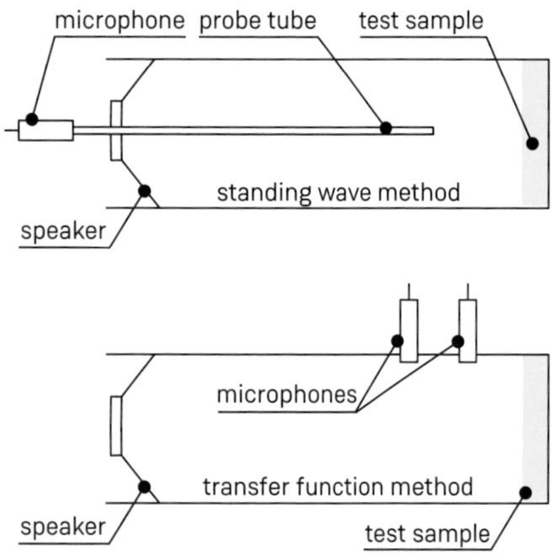

Diagrams of impedance tubes showing the standing wave measurement method and the transfer function measurement method. Impedance tubes are primarily used for prototyping and testing very small samples.

MEASURING ABSORPTION

There are a few methods for measuring acoustical absorption. Most commonly, materials are measured in a large Reverberation Chamber although for prototyping purposes, small samples of material can be measured in an Impedance Tube. Both measurement methods are defined in international standards.

Reverberation Chambers are so named because they have long reverberation times due to their large size and highly sound-reflective interior surfaces. The walls of the chamber are designed to provide sound isolation from exterior noise sources and are often composed of painted concrete or metal panels. Within the chamber, large reflective surfaces are suspended from the ceiling or mounted to the walls to scatter sound in various angles in order to create a diffuse sound field, so that sound waves strike the material test sample from many different directions. The dimensions of the reverberation chamber will limit the size of the specimen that can be measured as well as limit the lowest frequency that can be accurately measured.

To measure the absorption coefficients using the Reverberation Chamber (or Room Method), a large sample (or many samples) of material is installed in the chamber. The addition of this material will cause the reverberation time in the chamber to drop. The absorption coefficients of the material can then be calculated by comparing the measured reverberation times before and after the material was installed.

Samples are installed on the floor, walls, or on the ceiling of the chamber according to standardized mounting methods described in detail a following section. The sound absorption coefficient of a material measured in the chamber should not be considered a constant of the material. The absorption coefficients can change significantly depending upon the how the material is installed, the dimensions of the material, and how the material is distributed in the space. All of these factors should be considered when reviewing sound absorption data. Ideally, materials should be measured in a manner that most closely aligns with the way in which they will be installed in the field.

For prototyping purposes, material can also be measured in a small tube, using the standing wave or transfer function methods. With the standing wave tube, a loudspeaker generates a plane wave inside a tube which then reflects from the material sample at the opposite end of the tube. The impedance of the material under test will alter how sound is reflected and by measuring the standing wave within the tube using a moveable microphone probe, it is possible to calculate the absorption coefficient for sound at normal incidence – in other words, sound arriving perpendicular to the material surface. The standing wave measurement method can be time-consuming because absorption is measured one frequency at a time, however, it is a relatively simple and reliable process.

The Transfer Function Method uses two stationary microphone positions within the tube. When the measurements at both microphone positions are compared, formulas allow for the absorption coefficients and impedance to be calculated across a range of frequencies with only a few brief measurements. While this method is slightly more complex than the standing wave method, it is less time-consuming. Additionally, this method can be expanded upon to also measure sound transmission through a material.

Tube measurements are suited for prototyping and measuring small samples of material and commonly used to validate software prediction models for porous materials. A small sample, only a few centimetres in diameter, is necessary and the measurement apparatus itself is relatively small. The frequency range of the measurements is limited by the dimensions of the tube and multiple tubes are often used to cover different frequency ranges.

Because tube measurements only measure sound arriving perpendicular to the surface, they may not be representative of how materials will absorb at all other angles of arrival. In most real world applications, sound strikes a material's surface from many different angles. At a given frequency, the absorption coefficient of a material can vary significantly according to a sound's angle of incidence. For this reason, the reverberation chamber is preferred because it provides a diffuse field of sound arriving at random angles of incidence.

Another issue with tube measurements is that the absorption measured from a small sample is not always representative of the behaviour of a larger sample of material. As an example, while it would be possible to measure the absorption of a small piece of fabric using a tube, it would not be possible to measure that fabric as a drape with folds spaced 1 metre away from a wall. For such an application, the reverberation chamber method would be necessary.

ABSORPTION METRICS

All materials absorb sound to some degree, which can be measured in terms of a sound absorption coefficient, represented as alpha (α), which is the fraction or percentage of sound absorbed on a scale between 0.0 (reflective) and 1.0 (totally absorptive).

Absorption performance will vary significantly over the range of audible frequencies and tests commonly measure at the 18 different 1/3-octave band frequencies from 100 Hz to 5000 Hz. This data is then used to classify materials according to several metrics described here.

Single number metrics are useful for generalizations but there are drawbacks whenever representing complex data with a single number. For example, there may be instances where one octave band is very absorptive but the other bands are not. Averaging this data can result in a misleading indication of how the material actually absorbs. It is always recommended to consult laboratory test reports and consider the absorption coefficients across the range of frequencies.

For some materials, test reports can indicate absorption coefficients greater than 1. It is not actually possible for a material to absorb more than 100 per cent. The primary reason this anomaly occurs is due to the material thickness and edges, which are not accounted for when calculating the material's surface area. So for example, a panel that is 1200 mm x 600 mm x 100 mm has a stated surface area of 0.72m² in the measurement procedure, which considers only the plan-view surface area. However, when the edges are included in the actual surface area exposed to the room, it is actually 1.08m², which is a 50 per cent increase in surface area. Even when the edges are included in the total surface area, values slightly greater than 1.0 are still possible due to diffraction effects at the material's edges and corners.

Noise Reduction Coefficient [NRC] is a single number metric, defined in ASTM C423, which is the average of the absorption coefficients at the frequencies of 250, 500, 1000, 2000 Hz. This average is rounded to the nearest multiple of 0.05. The frequency range used to calculate NRC is representative of the range of human speech and is therefore useful for applications involving speech. NRC is perhaps the most common absorption metric and has been in use for a very long time. It is useful for providing a quick and rough approximation of a material's absorptivity. Because it only averages down to 250 Hz, it is not a good indicator of how materials absorb at low frequencies.

Sound Absorption Average [SAA] is a single number rating also defined in ASTM C423. It is the average of the absorption coefficients for the twelve 1/3 octave bands from 200-2500 Hz; this average is rounded to the nearest 0.01 increment.

Practical Sound Absorption Coefficient [αp], defined in ISO 11654, is given in octave bands (centre frequencies of 125-4000 Hz) as the arithmetic mean of the three corresponding 1/3 octave sound absorption coefficients, with the value rounded to the nearest multiple of 0.05.

Weighted Sound Absorption Coefficient [αw], defined in ISO 11654, is calculated using the practical sound absorption coefficients [αp] for the octave bands 250-4000 Hz. A reference curve, defined in the standard, is shifted against the practical absorption coefficients until the sum of the unfavorable deviations is less than 0.10. An unfavorable deviation occurs at a particular frequency when the measured absorption coefficient is less than the value on the curve. The weighted sound absorption coefficient is then defined as the value of the shifted reference curve at 500 Hz. Because this single number rating does not convey enough information, 'shape indicators' are often provided. These shape indicators suggest the sound absorption coefficient at one or more frequencies is considerably higher than the values of the shifted reference curve. [H] will indicate excess absorption at high frequencies, [M] for mid-frequencies, and [L] for low frequencies.

Absorption Class [A-E] , also defined in ISO 11654, categorizes absorption performance into five classes, where Class A has the highest performance. The curve from the weighted sound

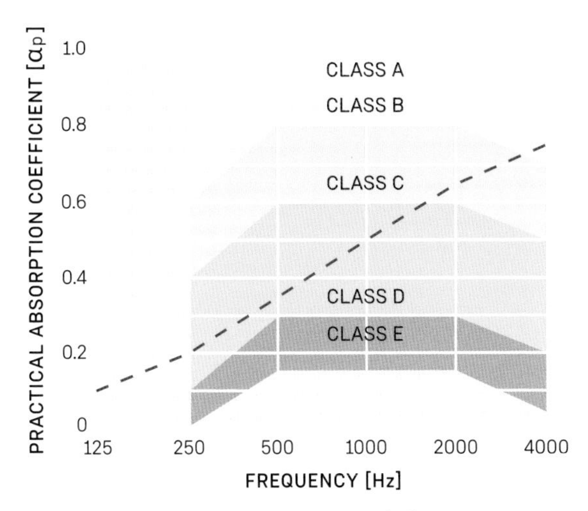

The practical absorption coefficients [αp] from the example table have been plotted (dashed line) against the Absorption Class contours. This example material, with a weighted sound absorption coefficient [αw] of 0.4 [H] is rated Class D. The [H] indicator suggests the material is more absorptive at high frequencies.

absorption coefficient [αw] procedure is used to define the five classes.

A **Sabin** is a unit of sound absorption equivalent to 1 ft^2 of a perfectly absorptive surface; a metric sabin is 1 m^2 of perfectly absorptive surface. For three-dimensional objects, such as suspended absorbers, the sound absorption performance is often given in terms of the number of sabins of absorption provided by each panel or object. The number of sabins per unit is equal to the total surface area of the unit that is exposed to sound multiplied by the absorption coefficient of that material.

1/3 Octave Band Centre Frequency [Hz]	Measured Sound Absorption Coefficient [α]	Practical Sound Absorption Coefficient [αp]
100	0.05	
125	0.08	0.10
160	0.14	
200	0.17	
250	0.20	0.20
315	0.24	
400	0.30	
500	0.39	0.35
630	0.41	
800	0.46	
1000	0.50	0.50
1250	0.43	
1600	0.61	
2000	0.64	0.65
2500	0.67	
3150	0.71	
4000	0.75	0.75
5000	0.79	
Weighted Sound Absorption Coefficient [αw]		0.40 [H]
Noise Reduction Coefficient [NRC]		0.45
Sound Absorption Average [SAA]		0.42
Absorption Class		D

This sample table provides the measured absorption coefficients of a material in a reverberation chamber. Various metrics based upon the measured 1/3-octave band data have been calculated (NRC, SAA αw, αp, Absorption Class). This type of data would be provided by any accredited testing laboratory. The data suggests the material under test absorbs well at very high frequencies but is a poor absorber at low frequencies. This is also underscored by the weighted sound absorption coefficient [αw], which has a shape indicator of [H] to suggest excess absorption at high frequencies.

MATERIAL MOUNTING

Whenever absorption data for a material is given, such as an NRC rating, the mounting method and depth of mounting that were used when testing the material should always be indicated in order for the absorption data to be meaningful. If this information is not indicated, the reader should request this from the manufacturer or sales associate. The way in which a material is mounted will have a significant impact on the absorption performance.

The figures shown here provide the common material mounting methods when tested in a reverberation chamber, as defined in the standard ASTM E795. In most cases, the test materials are installed or placed on the floor of the reverberation chamber, as this is the most practical location. For this reason, the figures shown may appear upside-down for materials such as suspended ceiling tiles (Type E mounting), which would normally be suspended from the ceiling above.

Types C, D, E, and G are further designated with numerical suffixes to indicate the mounting depth in millimetres between the specimen and the chamber floor, wall, or ceiling. For example, E-400 is a common mounting configuration for ceiling tiles, which indicates a mounting depth of 400 mm. For Type G, used to test shades or blinds, the mounting distance is given from the centreline of the hangers to the chamber wall.

Type A: SPECIMEN LAID DIRECTLY AGAINST THE CHAMBER.
Type B: SPECIMEN CEMENTED TO GYPSUM BOARD AND LAID DIRECTLY AGAINST THE CHAMBER.
Type C: SPECIMEN IS COMPRISED OF ABSORPTIVE MATERIAL BEHIND A PERFORATED, EXPANDED, OPEN FACING, OR OTHER POROUS MATERIAL.
Type D: SPECIMEN IS MOUNTED ON WOOD FURRING STRIPS.
Type E: SPECIMEN IS MOUNTED WITH AN AIR SPACE BEHIND IT.
Type F: SPECIMEN IS MOUNTED WITH AN AIR SPACE BEHIND IT WITH SPACERS THAT ARE INTEGRAL TO THE PRODUCT DESIGN.
Type G: SPECIMEN IS DRAPERY, WINDOW SHADE, OR BLIND HUNG PARALLEL TO THE CHAMBER SURFACE.
Type H: SPECIMEN IS A DRAPERY SUSPENDED AWAY FROM ANY VERTICAL SURFACE.
Type I: SPECIMEN IS SPRAY- OR TROWEL- APPLIED MATERIAL ON AN ACOUSTICALLY HARD SUBSTRATE.
Type J: SPECIMEN IS A SOUND ABSORBING UNIT OR SET OF UNITS, SUCH AS A CLOUD OR BAFFLE OR SIMILAR.
Type K: SPECIMEN IS AN OFFICE SCREEN, SUCH AS A CUBICLE WALL.
Type L: SPECIMEN IS CONCRETE-BLOCK OR BLOCK-LIKE AND ASSEMBLED USING MORTAR.
Type M: SPECIMEN IS THEATRE SEATS, ARRANGED IN A PATTERN TO SIMULATE ACTUAL INSTALLATION.

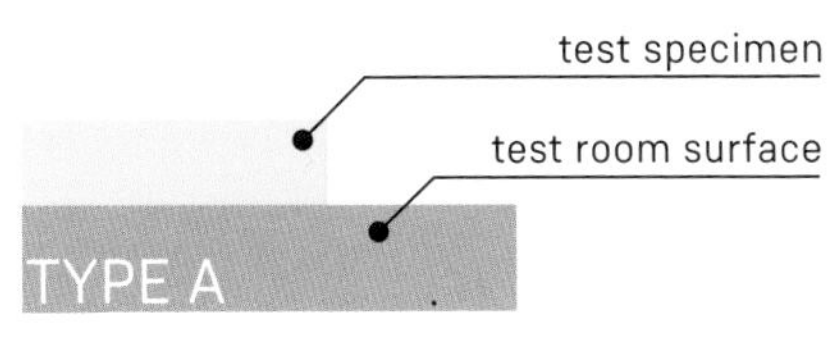
test specimen
test room surface
TYPE A

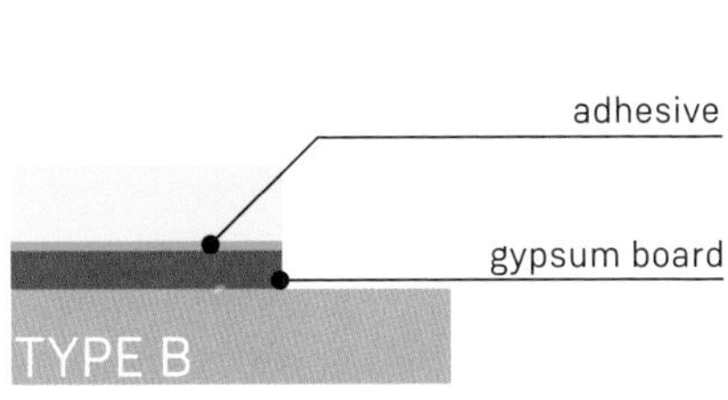
adhesive
gypsum board
TYPE B

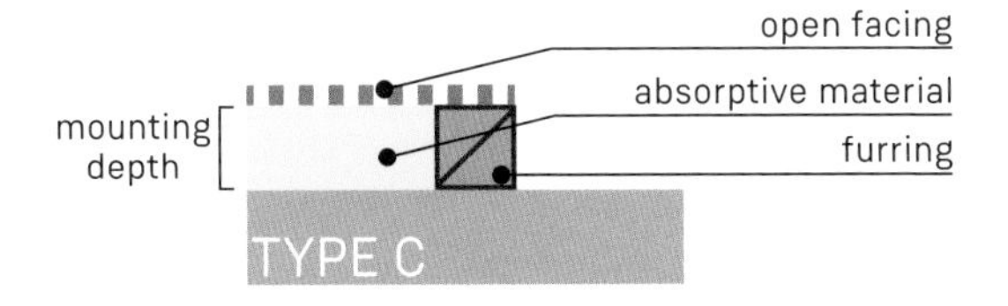
open facing
absorptive material
furring
mounting depth
TYPE C

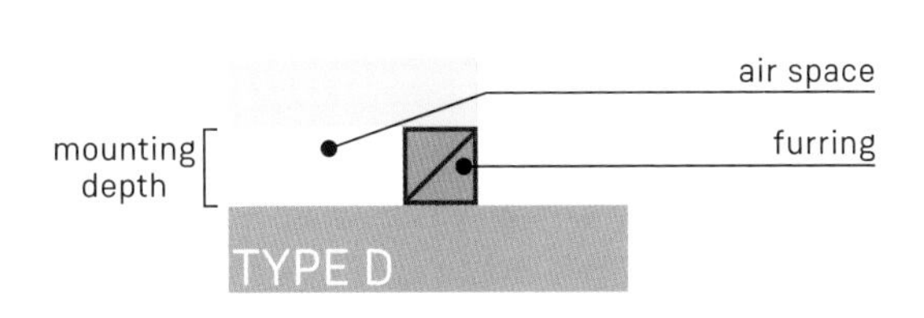
air space
furring
mounting depth
TYPE D

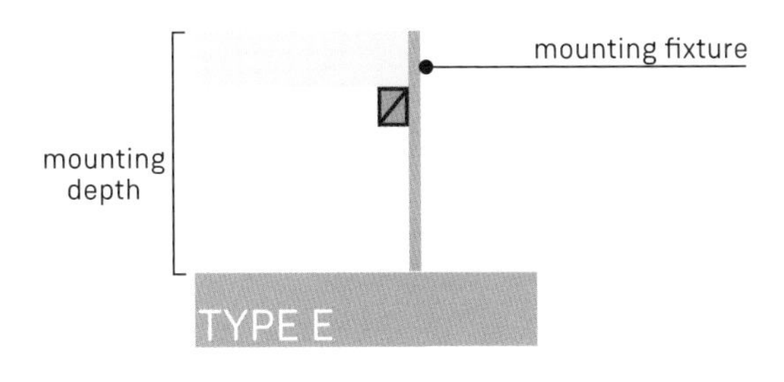
mounting fixture
mounting depth
TYPE E

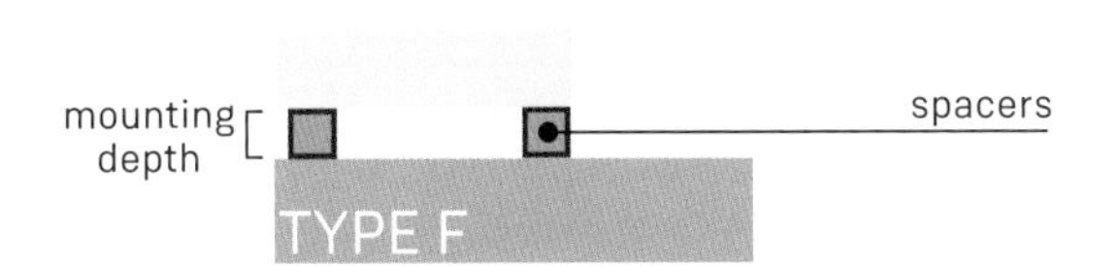
spacers
mounting depth
TYPE F

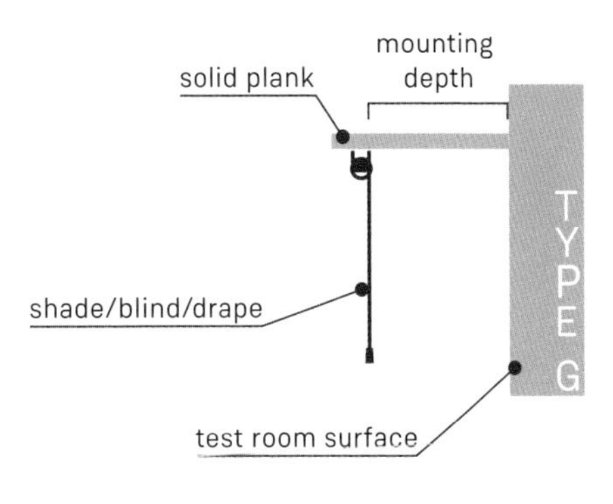
mounting depth
solid plank
shade/blind/drape
test room surface
TYPE G

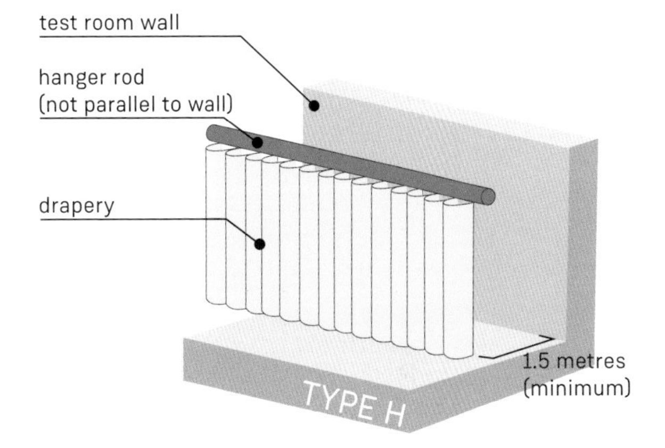
test room wall
hanger rod
(not parallel to wall)
drapery
1.5 metres
(minimum)
TYPE H

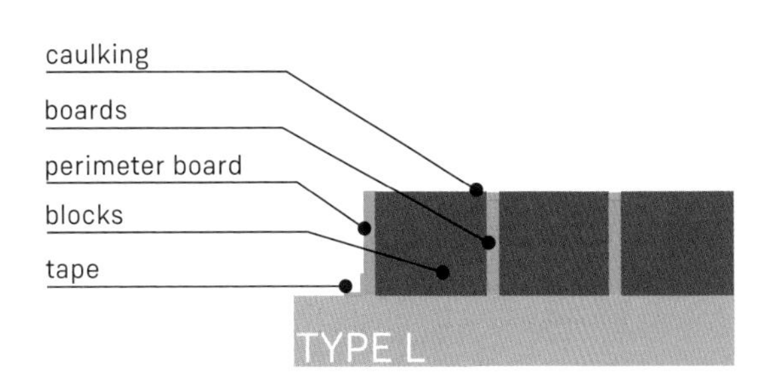
caulking
boards
perimeter board
blocks
tape
TYPE L

REVERBERATION TIME

Reverberation is the persistence of sound in a space. The metric 'reverberation time' (RT) is defined as the amount of time it takes for the sound pressure level to decrease by 60 dB after the sound source has stopped.

Reverberation time is related to the total volume of the room and the total absorption in the room, which was discovered at the beginning of the 20th century by Wallace Clement Sabine. The Sabine equation is still useful to this day for making rough predictions:

$$RT = \frac{k * V}{A}$$

RT = reverberation time (seconds)
k = constant (0.161 for metric; 0.049 for imperial)
V = room volume (m^3 or ft^3)
A = total surface absorption — the sum of the surface area of each material (m^2 or ft^2) multiplied by the material's absorption coefficient (α)

In order to calculate ***A***, the total surface absorption, the absorption coefficients and surface area of every material within the room must be known. For each material, its surface area is multiplied with its absorption coefficient and the sum of all materials comprises ***A***. An example calculation is given on the following pages.

The absorptivity of a material will vary according to frequency. For this reason, it is typical for the reverberation time to be calculated for each octave band from 125-4000 Hz. When an RT is specified without citing a specific octave band frequency, it is generally assumed to be the RT at 500 Hz.

The Sabine formula provides a basic yet reasonable approximation of reverberation time and demonstrates the fundamental relationship between material, volume, and reverberation.
It should be noted that the Sabine formula assumes a diffuse field, and rooms cannot be considered diffuse where acoustical treatments are concentrated to a single area; where rooms have parallel, flat reflective surfaces; rooms with reflective domes or concave walls; or rooms which have one dimension that is disproportionately different from the other two. For these reasons, the results predicted by the Sabine equation may be subject to inaccuracies.

REVERBERATION TIME CRITERIA

Ideal reverberation times will depend upon the type of use, the occupant's needs and expectations, and the size of the space. The figure shown below provides target reverberation times (at 500 Hz) for a variety of different uses. Spaces requiring speech intelligibility generally demand short reverberation times (1 second or less), whereas spaces for unamplified music performance will demand longer reverberation times to provide blending and support of the music - the length of reverberation will depend upon the room size and type of music. Opera and musical theatre will require shorter reverberation times compared to purely orchestral music because speech clarity is an important component of the musical experience.

Amplified music, such as dance music, may demand short reverberation times to provide articulation of beats. With amplified sound systems, whenever reverberation is desired, it is often artificially added into the mix over the loudspeakers and generally does not require natural reverberation from the room.

In terms of performance spaces, the ideal reverberation time can be a matter of taste that evolves over time. Reverberation times are not usually flat across all frequencies — longer reverberation times are often preferable at lower frequencies. Achieving longer reverberation times at low frequencies can be challenging because many common building materials act as panel absorbers, reducing sound below 250 Hz. Gypsum board attached to studs, for example, can absorb 20-30 per cent of sound at 125 Hz. On the other hand, in large performance spaces, maintaining high-frequency reverberation times can be challenging because the audience is absorptive and even the air itself, in very large rooms, begins to absorb sound at very high frequencies.

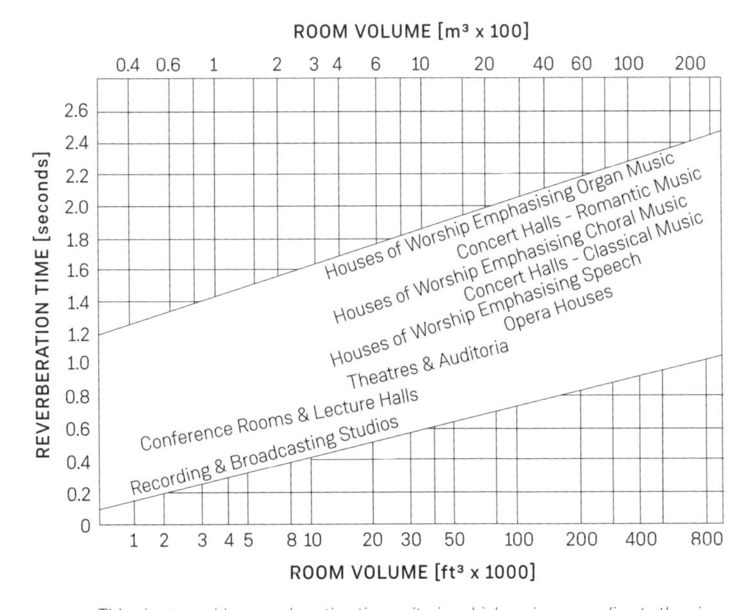

This chart provides reverberation time criteria, which varies according to the size of the room and the type of use. Whenever criteria are not given for each octave band, then it is common practice to reference the reverberation time at 500 Hz.

EXAMPLE RT CALCULATION

An example reverberation time (RT) calculation using the Sabine equation is presented here for two simple scenarios: a highly reverberant room (gypsum, concrete and glass finishes) and the same room with fibreglass added to the ceiling and thin carpet added to the floor. The room is 5 metres long, 4 metres wide, and 3 metres tall, resulting in a total room volume of 60 m^3.

For each room, tables are provided with the absorption coefficients for each surface material in octave bands. The surface area of the material is then multiplied by the absorption coefficient and this is summed together to give the total absorption within the room. In the last row, the RT is then calculated for each octave band.

If this room were being designed as a conference room, the criteria chart on the preceding page would suggest that a room of 60 m^3 should have a reverberation time of around 0.5 seconds at 500 Hz. The untreated room is indicating a reverberation time of 2.9 seconds at 500 Hz. Once fibreglass and thin carpet are added to the room, the reverberation time drops significantly. Using the formula presented in the preceding section, it is also possible to estimate the anticipated noise reduction due to the addition of absorptive materials. This formula compares the total absorption of the before and after conditions. For example, looking at the 500 Hz octave band, the before condition had a total sabins value of 3.4 and after adding absorption that value increased to 24.2. The reverberant noise reduction is anticipated to be 8.5 dB.

$$\text{noise reduction [dB]} = 10 * \log \frac{24.2}{3.4} = 8.5 \text{ dB}$$

It is interesting to note that the untreated room has very short reverberation times at low frequencies (less than 1 second at 125 Hz). This occurs because gypsum and glass are somewhat absorptive at 125 Hz.

A table of the absorption coefficients for common building materials is provided in the Appendix on page 282.

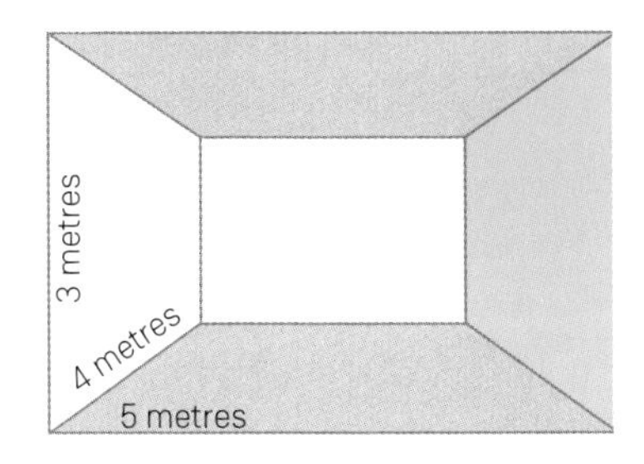

An untreated, 60 m³ reverberant room with concrete floor and ceiling, 3 gypsum walls, and 1 glass wall. Using the Sabine equation, the reverberation time at 500 Hz is 2.9 seconds.

Surface	Material	Surface Area		125 Hz	250 Hz	500 Hz	1000 Hz	2000 Hz	4000 Hz
Wall 1	Glass	12 m²	α	0.18	0.06	0.04	0.03	0.02	0.02
			S * α	2.16	0.72	0.48	0.36	0.24	0.24
Wall 2	Gypsum Board	15 m²	α	0.29	0.10	0.05	0.04	0.07	0.09
			S * α	4.35	1.50	0.75	0.60	1.05	1.35
Wall 3	Gypsum Board	12 m²	α	0.29	0.10	0.05	0.04	0.07	0.09
			S * α	3.48	1.20	0.60	0.48	0.84	1.08
Wall 4	Gypsum Board	15 m²	α	0.29	0.10	0.05	0.04	0.07	0.09
			S * α	4.35	1.50	0.75	0.60	1.05	1.35
Floor	Concrete	20 m²	α	0.01	0.01	0.02	0.02	0.02	0.03
			S * α	0.20	0.20	0.40	0.40	0.40	0.60
Ceiling	Concrete	20 m²	α	0.01	0.01	0.02	0.02	0.02	0.03
			S * α	0.20	0.20	0.40	0.40	0.40	0.60
		Sum Total	S * α	14.7	5.3	3.4	2.8	4.0	5.2
Reverberation Time [RT] (0.161 * Volume) / (Sum of S*α)				**0.7**	**1.8**	**2.9**	**3.4**	**2.4**	**1.9**

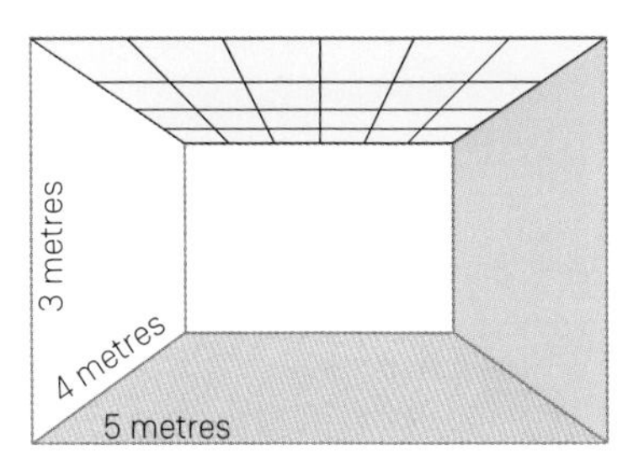

When fibreglass panels are mounted to the ceiling and thin carpet is installed on the floor of the same 60 m³ room, the reverberation time at 500 Hz drops to 0.4 seconds.

Surface	Material	Surface Area		125 Hz	250 Hz	500 Hz	1000 Hz	2000 Hz	4000 Hz
Wall 1	Glass	12 m²	α	0.18	0.06	0.04	0.03	0.02	0.02
			S * α	2.16	0.72	0.48	0.36	0.24	0.24
Wall 2	Gypsum Board	15 m²	α	0.29	0.10	0.05	0.04	0.07	0.09
			S * α	4.35	1.50	0.75	0.60	1.05	1.35
Wall 3	Gypsum Board	12 m²	α	0.29	0.10	0.05	0.04	0.07	0.09
			S * α	3.48	1.20	0.60	0.48	0.84	1.08
Wall 4	Gypsum Board	15 m²	α	0.29	0.10	0.05	0.04	0.07	0.09
			S * α	4.35	1.50	0.75	0.60	1.05	1.35
Floor	Thin Carpet	20 m²	α	0.01	0.05	0.10	0.20	0.45	0.65
			S * α	0.20	1.00	2.00	4.00	9.00	13.0
Ceiling	50 mm Fibreglass 16 kg/m² [A]	20 m²	α	0.22	0.67	0.98	1.02	0.98	1.00
			S * α	4.4	13.4	19.6	20.4	19.6	20.0
		Sum Total	S * α	18.9	19.3	24.2	26.4	31.8	37.0
Reverberation Time [RT] (0.161 * Volume) / (Sum of S*α)				**0.5**	**0.5**	**0.4**	**0.4**	**0.3**	**0.3**

MATERIAL INSTALLATION

The absorption performance of a material is dynamic in the sense that the performance can vary depending upon how the material is installed and where it is placed in relation to the sound sources. When installing absorptive material in a room, it is generally recommended to be as close as possible to the sound source; visible to the sound source (not hidden away); and uniformly distributed (not clustered).

In spaces where human activities are the sources of noise, the goal is to get the absorption closer to the humans. In spaces with horizontal dimensions greater than the vertical dimension, the ceiling is usually the best location for absorption because the walls are likely to be far away and no matter where an occupant moves within the room, the ceiling will be at consistently closer proximity than the walls. As well, the ceiling will provide a greater surface area for coverage and is less likely to be susceptible to dirt or impacts than the walls or floor. For spaces with a vertical dimension greater than the horizontal dimensions, it is often desirable to add absorption to the walls or to suspend additional materials from the ceiling so they are closer to the occupants.

Whenever absorptive materials are used, it is important that they remain exposed or visible to the sound source. If the material is tucked around a corner or hidden behind a whiteboard, projection screen, or cabinetry, it will not be an effective absorber. If the material is designed to double as a bulletin board, the addition of papers tacked to the face will also reduce the absorption performance.

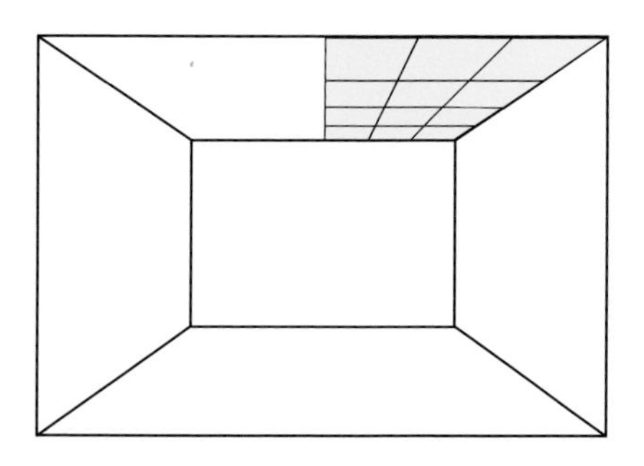

The top figure shows absorptive material distributed on the ceiling whereas the bottom figure shows the same quantity of material consolidated. Better acoustical performance can be achieved by distributing the material to provide uniform coverage throughout the space.

It is generally preferred to distribute material over an area rather than cluster or concentrate it to one part of a room. Distributing acoustical material can increase the efficiency of absorption - sometimes referred to as the area effect. Particularly when the panels are thick, sound will diffract around the edges of the panels, helping to create some diffusion. As well, the surface area of the exposed edges of each panel is greater when the material is spaced apart compared to being butt-jointed together.

When acoustical materials are mounted flush to the wall or ceiling, only one side of the material is exposed to the room. When that same material is suspended, both sides of the material are exposed (including the edges of the material) and the amount of absorption being provided by a single panel is effectively doubled. This type of design can result in cost savings because less material may be required to achieve the equivalent acoustical results of surface mounted panels. When panels are suspended vertically, often called Baffles, they may help block sound reflections between parallel walls. Panels suspended horizontally, often called Clouds or Islands, allow absorption to be suspended closer to the occupants and can be used to create the sense of a more intimate enclosure within a larger space. Suspended panels allow for absorptive materials to be installed in an existing space where the room finishes are already in place. However, adding suspended panels post-construction may require extensive coordination and planning to ensure the panels do not adversely impact lighting, HVAC circulation, or fire sprinkler coverage.

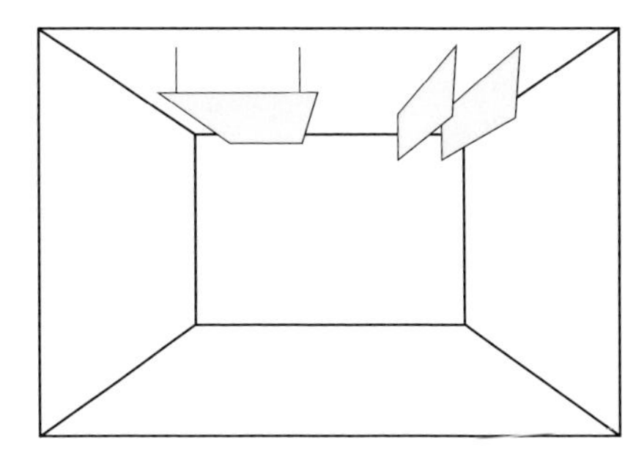

Absorptive material can be suspended horizontally (clouds or islands) or vertically (baffles). This type of installation exposes greater surface area of absorption to the room compared to a surface mounted panel.

MATERIAL CONSIDERATIONS

Acoustical materials are often required to serve more functions than simply sound absorption. When selecting a material, there are a number of performance requirements to consider, including visual appearance, cost, ease of installation, durability, cleanability, dimensional stability, mechanical strength, weight, and resistance to impacts and abrasion. In harsh environments, materials may be required to withstand high temperatures, moisture, or chemicals. In sensitive applications, such as clean rooms, food preparation areas, or medical environments, materials may be required to be non-porous and cleanable with high-pressure washing and disinfectants.

An acoustical material's light reflectance can be important when it's installed over a large interior surface such as a ceiling. Light reflectance may be expressed as a decimal fraction of the incident light that is reflected from the surface of the material. A high light reflectance will help provide efficiency for both daylight and artificial lighting and reduce the luminance ratio, or difference between light and dark surfaces, within a room. Light reflectance can be reduced due to dirt, so the cleanability or paintability of the surface may be another factor to consider when selecting an acoustical material.

The ability of an absorbing material to block sound may be another consideration. This is often measured in terms of transmission loss. One of the most common metrics for describing the sound transmission through walls and floors is the STC rating. Ceiling tiles are commonly rated with a Ceiling Attenuation Class (CAC) rating that indicates the degree to which sound passes through the tile into the ceiling plenum. Since many porous materials do not block sound, they are often combined with other materials to create a composite that will both absorb and block sound. As an example, quilted absorbers are often equipped with a layer of mass-loaded vinyl so that sound is both absorbed by the quilted material and blocked by the vinyl barrier.

A material's reaction to fire is one of the most important performance requirements, whether used in transportation, in equipment enclosures, or as an architectural finish. Materials are tested in a number of ways to evaluate their contribution to the early stages of a fire, namely: ignitability, flame spread, heat release, smoke production, and occurrence of flaming droplets or particles. Complete building elements, such as walls, floor/ceiling assemblies, and doors, are tested in terms of their fire resistance or endurance to a fully developed fire. Material flammability requirements will vary according to the project application and each jurisdiction may have its own particular code requirements.

Sustainability is one of the newer performance requirements to emerge. Traditionally, building codes have focused on the health, safety and well-being of occupants. Concerns about green and sustainable building practices began to emerge in the 1990s. Codes did not always address these concerns and independent rating systems such as Leadership in Energy and Environmental Design (LEED) and Green Globes emerged, which have served as the foundation for integration into a variety of code requirements and design standards.

Many of these standards will place requirements upon a material's use of recycled content; the ability to recycle or reuse the material when it is de-installed; the sourcing of raw materials from sustainable sources; the toxicity of materials including paints, coatings, and binders; the type and amount of energy used in the production process; and the distance to transport the material to the job site.

Life Cycle Assessments (LCAs) were developed to provide a more comprehensive picture of sustainable products, from the extraction of raw materials to what happens at the end of the product's life. This has also been coined as a 'Cradle-to-Cradle' system. The advantage of this approach is that synthetic materials are not discounted; in many cases, they actually turn out to have less of an impact than those derived from natural resources.

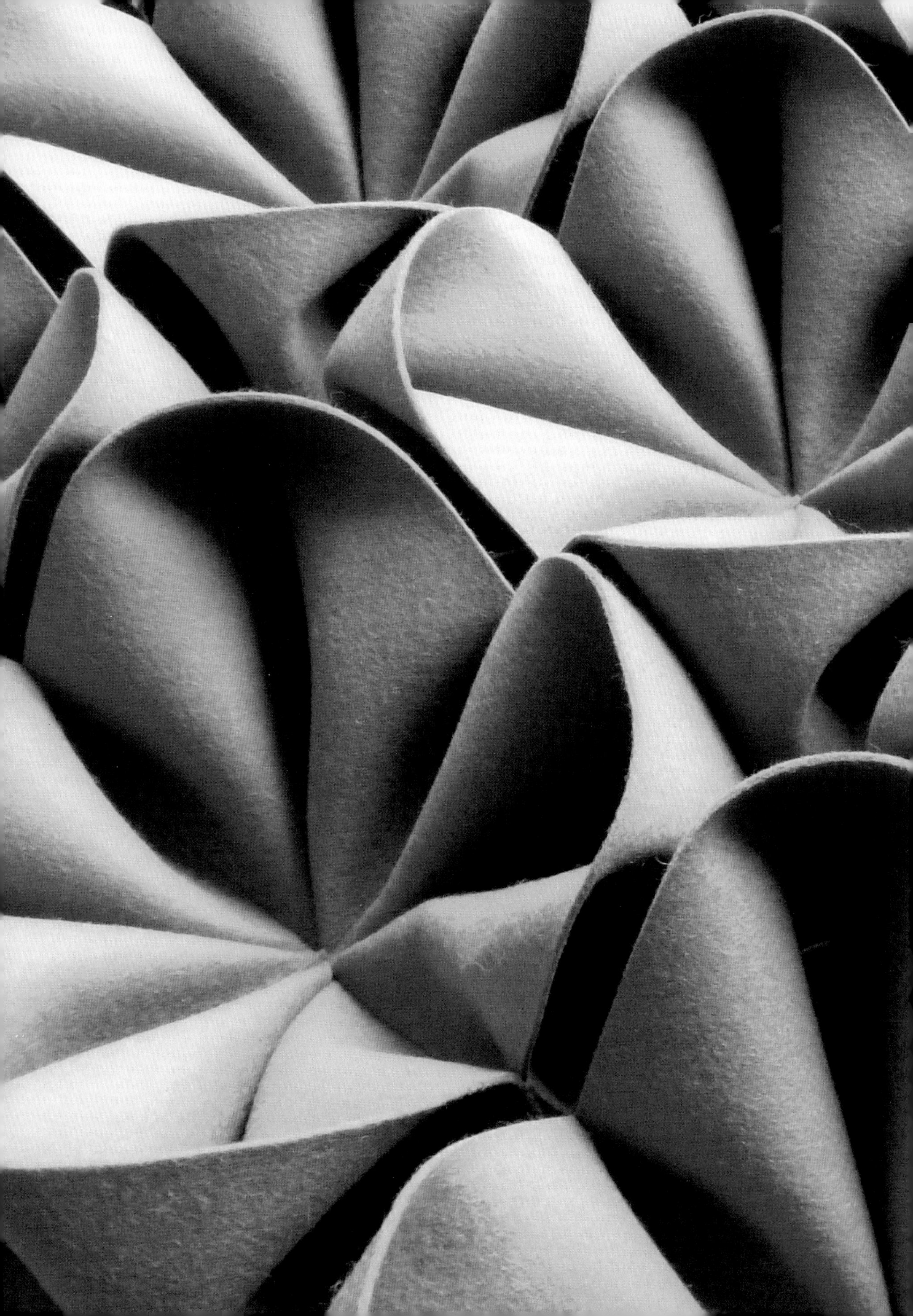

POROUS

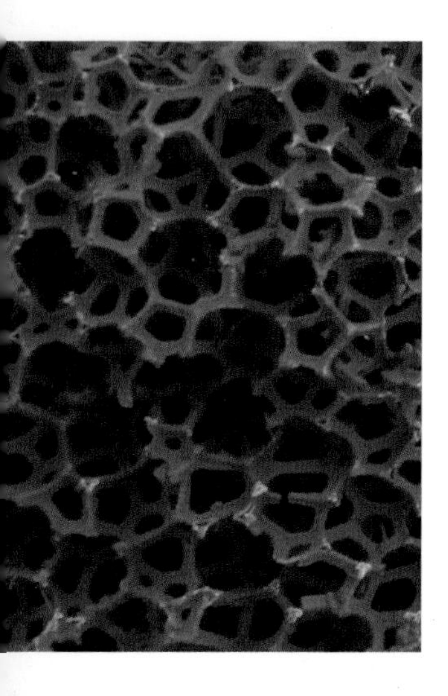

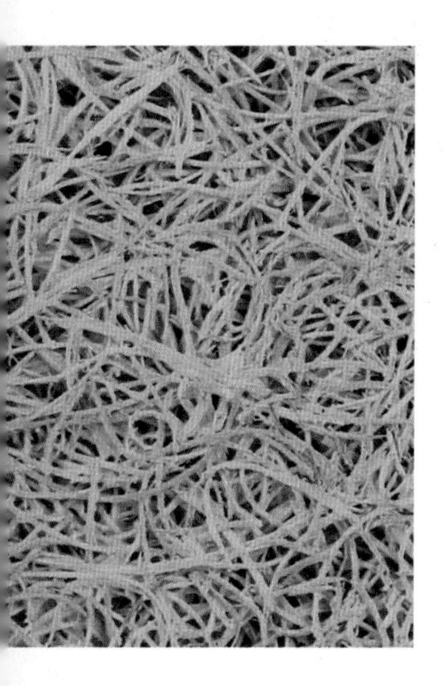

OVERVIEW

Porous absorbers are the most common type of acoustically absorptive material - soft, fuzzy, spongy materials like fibreglass, foam, and heavy textiles. With porous materials, sound absorption occurs due to interconnected pores creating viscous effects that cause acoustic energy to be dissipated as heat. In order for this to occur, the pores must be interconnected throughout the material and these pores must be open or exposed to the environment.

Porous absorbers fall into the three categories according to their porous micro-structure: cellular, fibrous, and granular. Foams are an excellent example of cellular materials. Fibreglass is an example of fibrous material, where air pockets are created between the overlaying fibre filaments. Granular materials create air pockets between tiny grains that are packed together — examples include some types of porous concrete, gravel, and soils.

The absorptivity of porous materials is primarily related to porosity and flow resistivity. Porosity can be defined in terms of the fractional amount of air volume within the material. It is the ratio of the total pore volume to the total volume of the material. The higher the porosity, generally, the better the absorptivity. Glass and stone wools can have a porosity of around 98 per cent.

Pores must be open and interconnected. Pores that are entirely isolated from their neighbours are called 'closed' pores or cells. They have an effect on some macroscopic properties of the material such as its bulk density, mechanical strength and thermal conductivity but closed pores do yield significant sound absorption. Closed cells do not count toward a material's total porosity.

Open pores that are not interconnected are considered 'blind'. Another important distinction is the difference between surface roughness and porosity — a rough surface is not considered porous unless it has irregularities that are deeper than they are wide.

Porous materials can also be categorized in terms of their flow resistance — a measure of the resistance experienced by air when it passes through the open pores in the material. If the flow resistivity is too high, then there is an impedance mismatch between the air and the material causing sound to reflect rather than absorb into the material. Flow resistivity is one of the most important parameters that determines the absorptivity of a material.

Other important factors for sound absorption include the pore shape, length and tortuosity. Pore shapes can have different surface areas, which can influence the thermal and viscous effects. Generally the more tortuous or complex the path of interconnected pores, the greater the absorption. The orientation of these paths can factor into how easily air can penetrate the material, which relates back to flow resistivity.

For fibrous materials, an important parameter is the diameter of the fibre as this relates to flow resistance. Synthetic fibres can be manufactured in a range of diameters and their composition tends to be very consistent compared to natural fibres, which may have a range of fibre diameters with irregular shapes. In general, natural fibres tend to have slightly larger diameters than some of the synthetic fibres that are commonly produced.

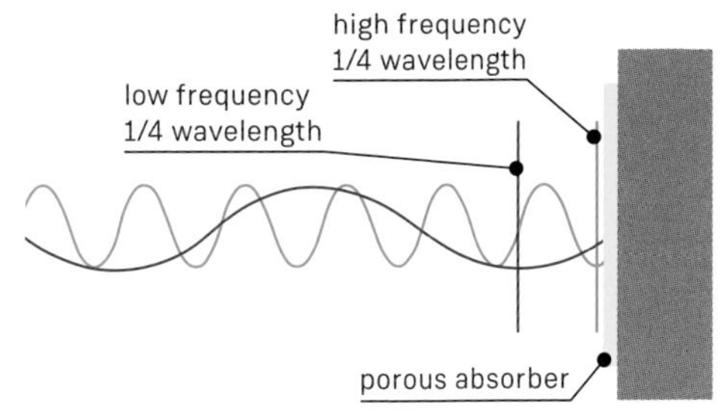

Sound absorption occurs when a porous material is 1/4 wavelength or more from the reflecting surface. Low frequencies have long wavelengths, requiring very thick materials.

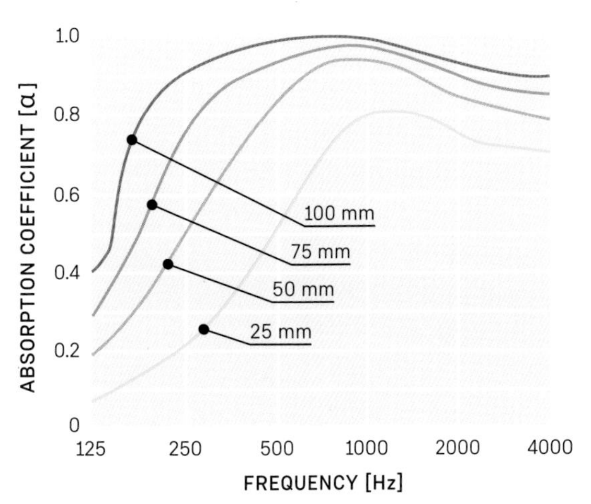

Sound absorption coefficients for fibreglass blankets (52 kg/m³ density) at four different thicknesses [A]- mounted directly to the wall.

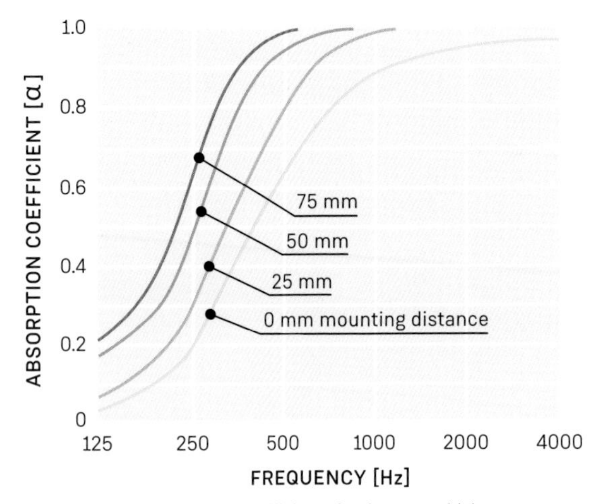

Sound absorption coefficients for the same thickness fibreglass panel when mounted at different distances from the wall.

POROUS THICKNESS

For porous materials, absorption increases at low frequencies as the material thickness increases. Porous materials absorb most effectively when placed where the particle velocity is high - typically 1/10 of the wavelength or more from the boundary surface (i.e. the wall,ceiling, or floor) to provide marginal absorption, 1/4 wavelength or more to provide significant absorption. The particle velocity directly at the boundary surface is nearly zero. For materials installed against the boundary surface, as the material thickness grows, the material's distance from the boundary surface also increases, resulting in better absorption for longer wavelengths or lower frequencies.

POROUS MOUNTING

Mounting porous material away from the boundary surface is one way of increasing the low frequency absorption without needing to increase the overall thickness of the material. As graphed, the same material can achieve significantly different absorption characteristics depending upon the mounting conditions. For this reason, it is always imperative, when reviewing absorption test reports, to consider how the material was mounted during the testing.

Porous absorbers are not often practical for absorbing very low frequencies (bass traps) because they need to be extremely thick or spaced very far from the room boundary (wall/ceiling/ floor). For example, at 90 Hz, the total wavelength is nearly 4 metres long. A porous material would need to be located approximately 1 metre from the boundary surface to achieve the 1/4 wavelength requirement to provide significant sound absorption.

Many believe that placing porous absorption in the corners of the room is an effective bass trap. However, while modes have maximum pressure in the corners, the particle velocity is low and so the porous absorption is not terribly effective. Low-frequency absorption can be addressed using panel and other resonant types of absorption. Multiple types of absorbers are sometimes needed to provide a balance of absorption across the frequency spectrum.

POROUS MATERIAL DENSITY

Porous materials may be manufactured in a variety of densities. For example, stone or glass wool can be fabricated in soft and flexible blankets or dense and rigid boards. Density can correlate with flow resistivity, so for extremely dense boards, it becomes increasingly difficult for sound to penetrate the material and the material can become reflective. When densities are extremely low, then the resistivity drops and the material may become less absorptive. For most glass or stone wool acoustical board products, there is not a huge difference in the absorption performance between the standard densities that are commonly available.

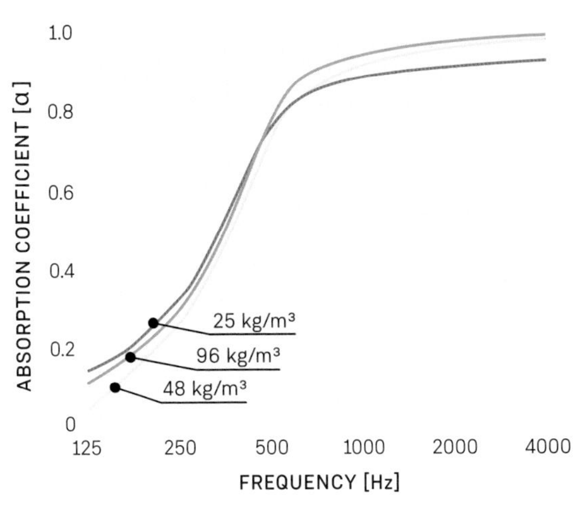

25 mm thick glass fibre board in three different densities mounted directly to the wall (A-mounted).

EFFECTS OF PAINT

Paint can reduce the absorption of porous materials because the paint may seal the pores or perforations and prohibit sound from penetrating the material. It is advisable to always check with the material manufacturer regarding maintenance, cleaning, and repair procedures. Some manufacturers may allow for a few layers of non-bridging water-based paints to be applied without significantly decreasing absorption.

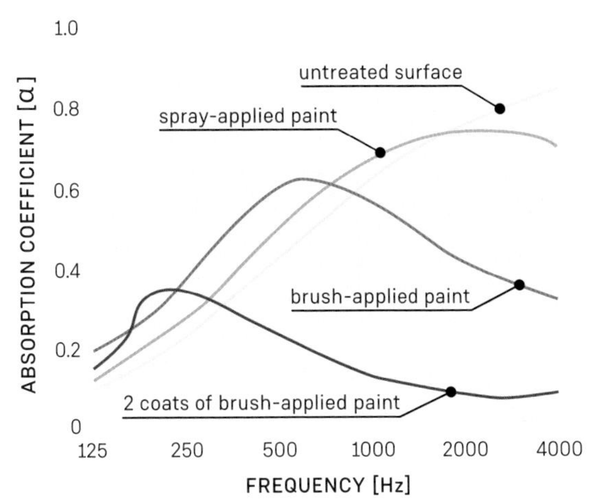

Examples of how paint can impact the absorptivity of a porous material.

POROUS ABSORPTION IN PARTITION CAVITIES

Porous absorbers are widely used in cavities of walls or floor/ceiling assemblies (typically in the form of batt) to improve the sound isolation performance of the partition. In some cases, these materials double as thermal insulators. The porous material is used to reduce the resonance within the air gap of the leaves of the partition. If the cavity resonance is not reduced by damping, then sound can more easily pass through the partition at the frequency range of the resonance. A variety of porous material can be used within the partition cavity for this resonance to be reduced.

VARIABLE ABSORBERS

In many circumstances it may be desirable to have a dynamic acoustic environment which can adapt to support various types of activities. Performance spaces, for example, may demand an acoustically deadened space for lectures and a livened space for unamplified music performances. Variable absorption allows for subtle adjustments to the room acoustics by varying the amount of absorptive material exposed to the room.

Draperies are perhaps the most universal and cost-effective form of variable absorber, allowing for gradual adjustments based upon the surface area of fabric exposed to the room, the amount fullness in the folds of the fabric, and the distance between the fabric and the reflecting surface. A number of other novel concepts for variable absorption are also presented here. Hinged panels can be fabricated such that the absorptive material folds inward to reveal reflective finishes. Rotating units can reveal angled or curved surfaces for diffuse reflections. Louvred or layered perforated panels create small adjustable apertures in front of the absorptive backing material.

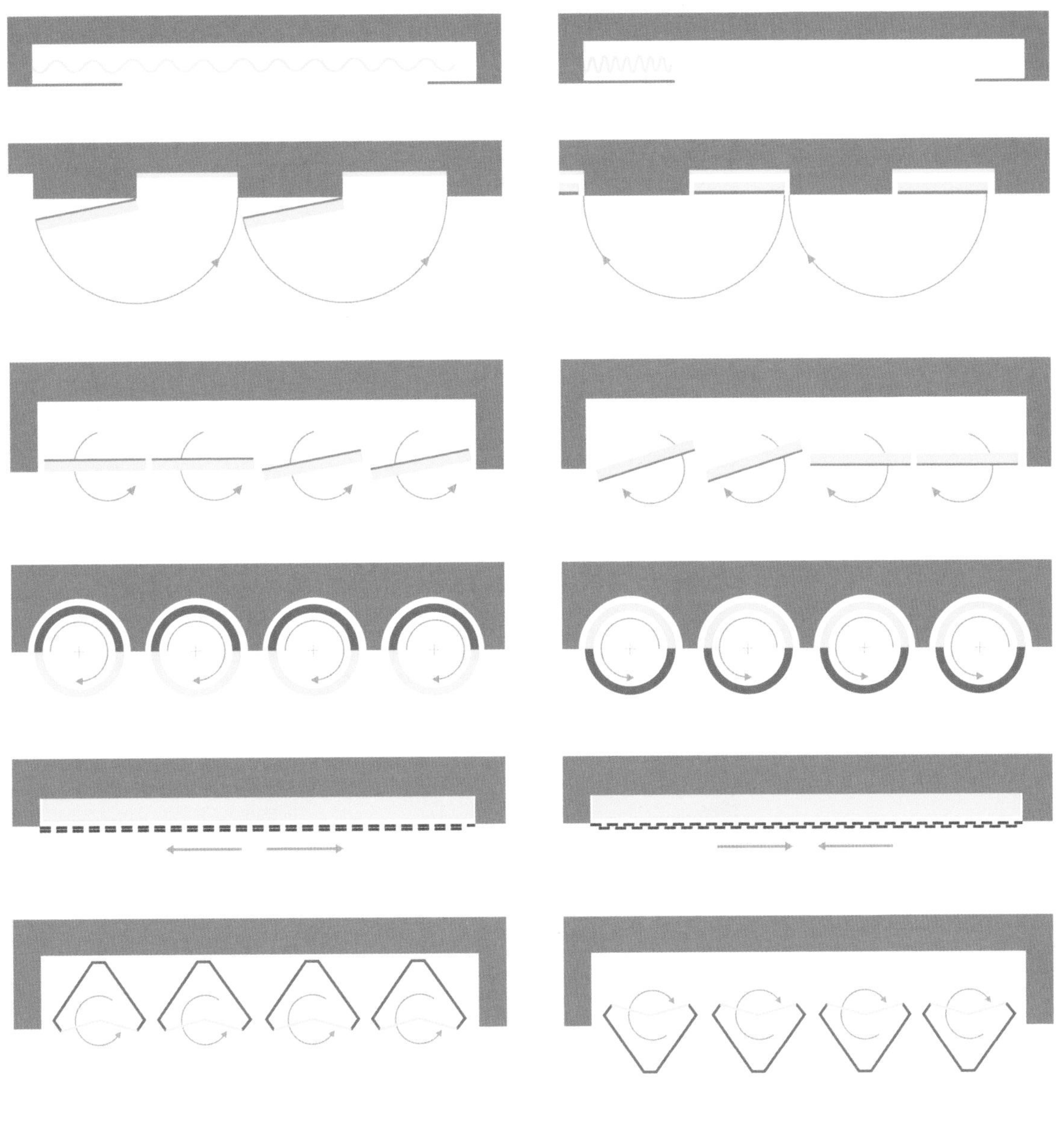
ABSORPTIVE
REFLECTIVE

Resonant Chamber, created by design firm rvtr (2011), is an interior envelope system based upon principles of origami to dynamically change the acoustic environment by varying the exposure of reflective (bamboo) or absorptive (perforated bamboo with PEPP absorptive backing) surfaces. The tessellated surface is kinetically controlled with linear actuators that expand or contract. Resonant Chamber not only reflects/absorbs sound but can produce sound through distributed loudspeakers.

CARPETS

Carpets are essentially porous type absorbers. They are efficient at absorbing high frequencies but due to limited thickness are usually ineffective absorbers at low frequencies. The amount of absorption, which can range between NRC 0.10 up to NRC 0.60, will depend upon factors such as the pile and weight of the carpet, the type and thickness of the underlayment, and whether or not the carpet has an impermeable backing.

Porous, permeable underlayments such as open-celled foam, open-celled sponge rubber, or old-fashioned felt hair materials are all acoustically absorptive. Whereas, closed-celled foams, closed-cell rubbers, or rubber-coated underlayments are not as absorptive. The more permeable the carpet and the underlayment, the more sound can penetrate and be absorbed. As with all porous type absorbers, the thicker the material, the greater the absorption.

Cut piles provide slightly more absorption than loop piles. With cut pile, sound absorption increases along with increases in pile height and pile weight. With loop pile, when pile height increases with the density held constant, the sound absorption improves; when the pile weight increases, with the pile height held constant, sound absorption rises only up to a certain level.

In addition to absorbing sound, carpets are also effective in reducing noise created by footfall, chair scrapes, and rolling carts or luggage being pulled across the floor.

It is uncommon for carpet manufacturers to conduct acoustical absorption tests on their products. For this reason, there is relatively limited data regarding the absorptivity of specific brands or configurations with underlayments. The data available is limited to a few detailed studies, some of which were conducted several decades ago. Understanding the underlying principles will allow one to estimate the absorption performance. Studies have shown a pretty close correlation between the NRC rating and the combined thickness of the carpet and underlyament.

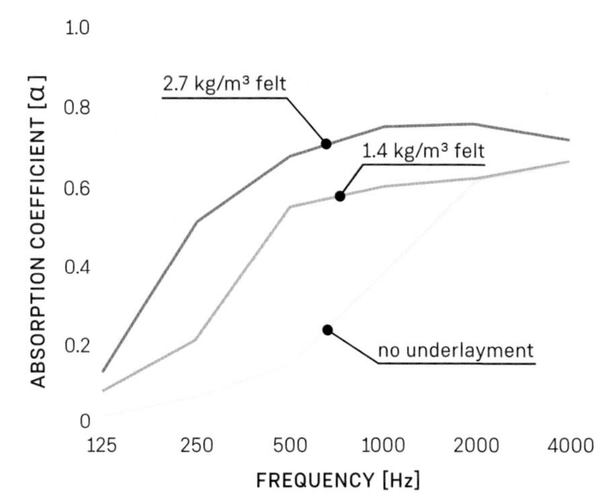

Absorption coefficients for the same carpet with different thicknesses of felt underlayments.

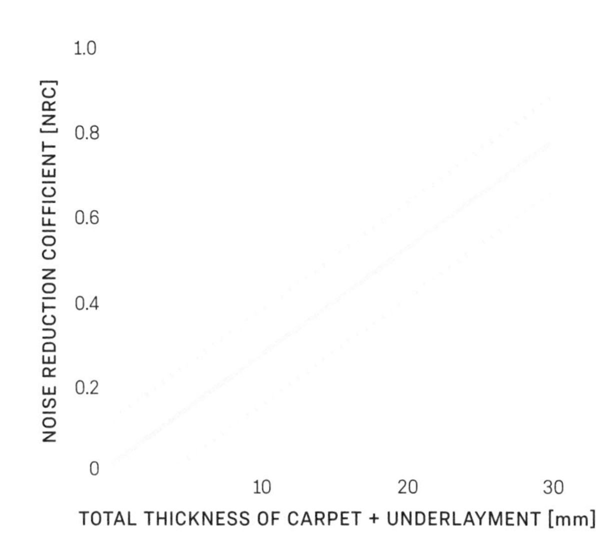

Average sound absorption (NRC) of carpeting as a function of total thickness of the carpet and underlayment based upon twenty-five test samples. Dashed lines represent the 95 per cent confidence limits.

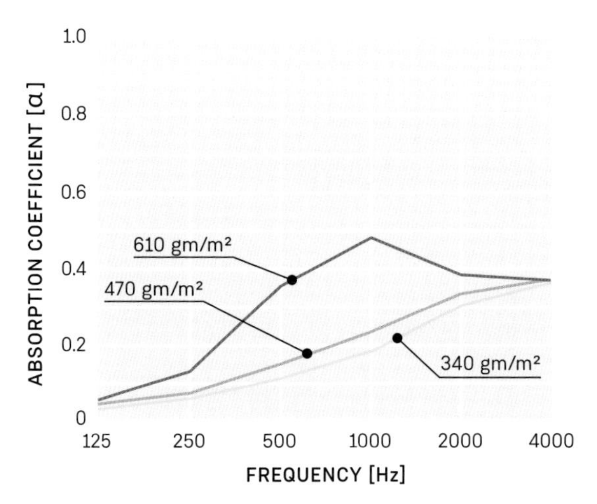

Absorption coefficients of velour drapery hung flat for three different weights of fabric.

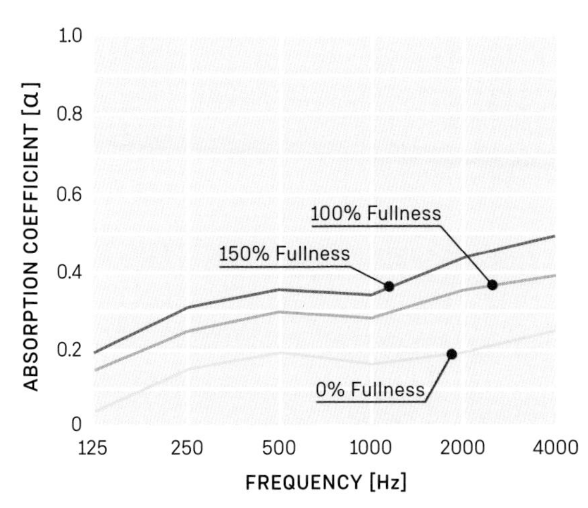

Absorption coefficients for the same drapery material when hung with different percentages of fullness.

DRAPERY/CURTAINS

Many textile curtains or drapes may be classified as porous absorbers. There are a few factors which impact the sound absorption performance of drapery; namely, the airflow resistance of the textile, the mounting distance of the fabric to the backing surface (wall or window), and the drapery fullness. The heavier the fabric, the greater the sound absorption. Heavy velour drapesm for example, can provide good absorption, while traditional lightweight sheer fabrics provide very little absorption.

Fullness is a term that describes the amount of extra fabric used to hang a drape. Fullness is typically provided in the horizontal dimension of the fabric but in some cases can occur in the vertical dimension. A drape sewn flat and installed flat is said to have 0 per cent fullness. Whereas a drape with 100 per cent fullness uses two times as much fabric as the horizontal dimension of the installed condition. A drape with 200 per cent fullness uses three times as much fabric as the horizontal dimension of the installed condition.

Fullness can be achieved through installation by gathering or rippling the drapery, or it may be manufactured into a drape through various pleating or shirring techniques. For the most part, the deeper the folds, the better the absorptivity because this results in more material surface area.

Spacing drapery away from the reflecting surface of the wall or window can also improve the absorption efficiency. Porous materials work well when they are approximately one quarter wavelength from the reflecting surface. If a drape has fullness, the folds will create varied distances away from the wall, broadening the absorption.

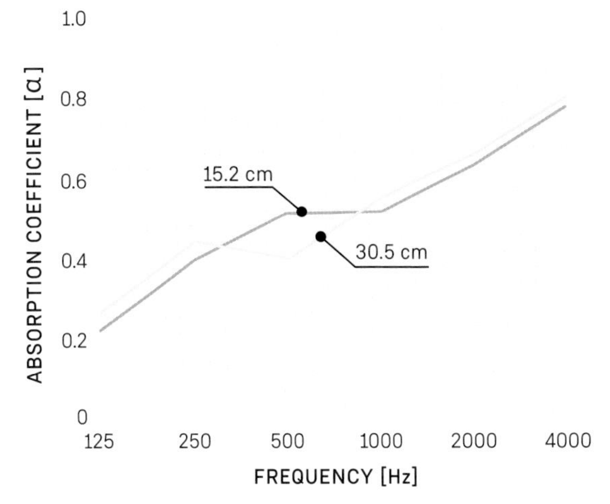

Absorption coefficients for the same drapery material (100 per cent fullness) when hung at two different distances from the wall.

50 mm thick board made of stone wool.

STONE & GLASS WOOL

Man-made Vitreous Fibres (MMVFs), such as stone and glass wool, have been in use for many decades and are the most widely used materials for sound absorption. They are low cost, durable, lightweight, easy to install, and may provide thermal insulation. These materials are incorporated into a variety of acoustical products such as ceiling tiles, acoustical plaster substrates, duct liners, and Helmholtz absorbers. The leading manufacturers and international brands are OwensCorning, Knauf, JohnsManville, Roxul, and Saint-Gobain, which is the parent company of CertainTeed and Isover.

Stone and glass wool is manufactured by spinning or drawing molten minerals into thin fibres, similar to cotton candy, which are finely coated with an adhesive that binds the fibres together. The material can be compressed into a variety of densities, and products are commonly available as loose fill materials, batts, blankets, and boards. The intertwined fibres create a network of tiny air pockets, resulting in a low-density and highly porous material.

Stone wool (rock wool or mineral wool) was first made in the mid 1800s in England and later produced commercially in Germany in the 1870s. The process involved blowing pressurised air across a flow of liquid iron slag, inspired by a type of naturally occurring lava called Pele's Hair that is created when droplets of ejected lava are elongated by the wind. Today's stone wool is typically made from rock such as basalt and iron-ore slag, a recycled by-product from iron manufacturing.

Glass wool (fibreglass) was invented and patented by Games Slayter in 1933, an employee of Owens-Illinois Glass, which later became Owens-Corning. Glass wool is made from some percentage of recycled glass content, such as from windows and bottles, as well as sand, soda ash, limestone, and other minerals.

For acoustic applications, stone and glass wool are comparable in terms of absorption performance. These materials can be fabricated in a variety of fibre diameters, thicknesses, and densities, which influence the acoustic performance. Material preferences can be driven by secondary factors such as thermal insulation performance, availability, weight, or fire resistance. While both stone and glass wool meet the highest fire ratings for building safety, the highest operating temperature for stone wool is 850° C; whereas glass wool is 230° C. Another distinguishing factor is that stone wool fibres are often coated with an oil-based spray that renders them hydrophobic — water beads on the surface but does not absorb into the material.

Health concerns regarding MMVFs emerged in the 1970s in Europe and the United States, prompting extensive study of potential carcinogenic risks of stone and glass wool. These concerns are sometimes confused with asbestos, which is a naturally occurring fibre that is a known carcinogen that is banned in most countries.

50 mm thick board made of glass wool.

In 2002, the International Agency for Research on Cancer, part of the World Health Organization, issued a comprehensive report on MMVFs and reclassified bio-soluble glass and stone wool fibres to category 3: 'not classifiable as to their carcinogenicity to humans.' The European Union reclassified stone wool in 2009 and, in 2011, the United States including California no longer required cancer warning labels on fibreglass and mineral wool products for acoustic and thermal insulation. The small fibre size of many stone and glass wool products may cause minor irritation to skin, eyes, and throat during handling, installation, or removal. This is due to mechanical action and not necessarily an allergic reaction. Manufacturers will provide safety guidelines for handling, which may involve wearing long sleeves, dust masks, or eye protection.

There are also health concerns regarding formaldehydes in building materials. The leading formaldehyde off-gassing building materials are plywood, particle board, and medium density fibre board. Some glass wool products have used formaldehyde as an ingredient in the binder that holds the fibres together and gives the insulation its shape and ability to recover from compression. There have typically only been trace levels of formaldehyde in traditional glass wool products, particularly when compared to many other building materials. In response to public concerns for safety and sustainability, many manufacturers are phasing out formaldehydes altogether. For example, EcoTouch insulation, produced by OwensCorning, is entirely formaldehyde-free, Greenguard Gold Certified and made with 99 per cent natural materials and a minimum of 58 per cent total recycled content (36 per cent post-consumer recycled content).

Samples of black acoustical board products, left to right: Knauf Black Acoustical Board, Isover SSP2, OwensCorning SelectSound.

BLACK ACOUSTIC BOARD/BLANKET

Stone or glass wool products are available from many manufacturers with a thin, matte black non-woven scrim fabric covering, which helps protect the fibres from impact, abrasion, dirt, and friability. This material is commonly used to line the inside of ducts for heating and air-conditioning. The material can also be installed directly to walls and ceilings using adhesive or impaling pins and is a very cost-effective solution for blacked-out spaces such as movie theatres, sound stages, and back-of-house performing arts spaces. The product is sometimes referred to as 'theatre board'. Since the material is matte black, it does not reflect light, and it is frequently used as the backing material behind perforated panels or screens, or in a ceiling plenum above a suspended grid, to visually hide the presence or appearance of acoustical material. As with any stone or glass wool product, it is manufactured in a variety of thicknesses and densities, in rigid or semi-rigid self-supporting boards and flexible blankets.

CEILING TILES

Suspended acoustical ceilings rose to popularity in the 1950s and continue to be a staple of contemporary construction and architecture for virtually any conceivable application. A typical dropped or suspended ceiling consists of a grid-work of metal channels in the shape of an upside-down 'T', suspended on wires from the overhead structure. These channels snap together in a regularly spaced pattern of cells which are each filled with lightweight ceiling tiles that simply lay into the grid.

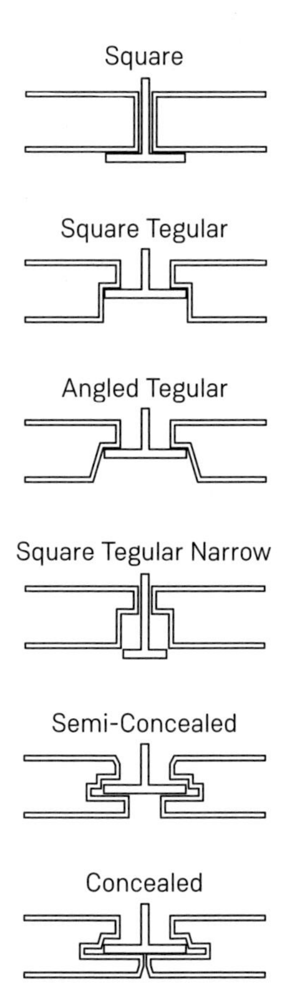

Common edge treatments for ceiling tiles. The way in which the tile edge interfaces with the T-bar will produce different visual effects. Edges may be designed so that the tile protrudes from the grid or conceals the grid entirely.

When the tiles lay on top of the T-bar, the grid is visually exposed. Some tile profiles allow for a semi- or fully-concealed grid, creating a more monolithic surface. The concealed grid causes the panels to interlock, which can make it difficult to access the plenum or replace the tile. These concealed grid ceilings may have a 'key panel' which can be removed, allowing the other panels to sequentially slide out of the grid. In the United States, the grid/tile sizes are typically 2 ft × 2 ft or 2 ft × 4 ft. In Europe, the cell size in the suspension grids is 600 × 600 mm, while the ceiling tiles are slightly smaller — 595 mm x 595 mm or 595 mm x 1195 mm. With this level of standardisation, one can easily calculate the dimensions of a room simply by counting the tiles in the ceiling.

The suspended grid ceiling tackles numerous building concerns with a singular, cost-effective, and modular system. Building services, such as HVAC, plumbing, and electrical, can be hidden from view yet remain easily accessible. The plenum created by the cavity behind the ceiling tiles can also be used for air circulation without requiring additional ductwork, resulting in some cost savings. The modularity of the system allows for lighting and HVAC to easily integrate within the grid and be easily reconfigured for changing floor plans.

In terms of acoustics, drop ceilings are capable of providing high levels of absorption with relatively thin materials because they are mounted with a large air cavity. Ceilings are often the best location within a room to install sound absorption because it is the surface least susceptible to impacts or damage. For many spaces, the ceiling surface area is greater than the surface area of the walls, resulting in a high degree of coverage and providing uniform coverage throughout the room. Ceiling tiles can also be

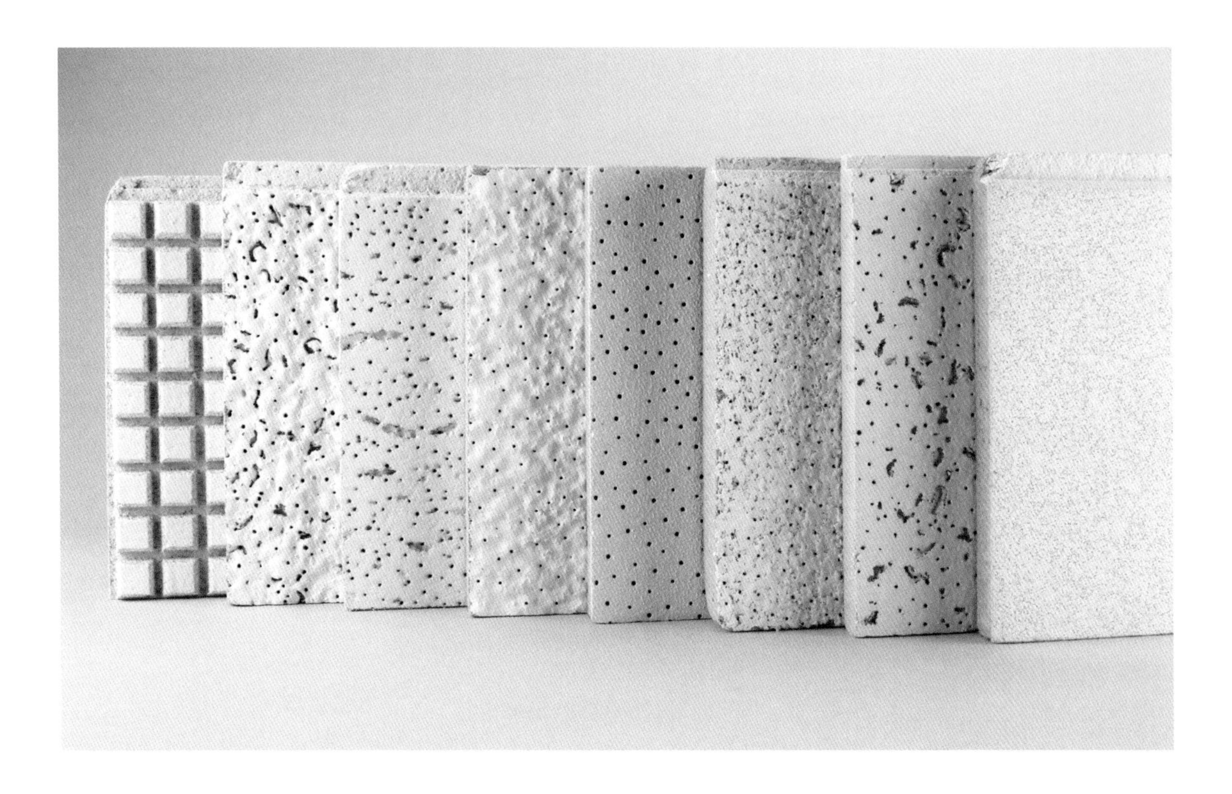

Armstrong Mineral Fibre Tiles, left to right: Graphis Cubic, Designer, Fissured, Armatuff, VL Perforated, Cirrus, Cortega, and Canyon.

used to block sound from mechanical ducts and equipment in the ceiling plenum, such as VAV boxes or fan coil units.

The main types of acoustical ceiling tiles are fibreglass, mineral wool, and mineral fibre. Fibreglass and mineral wool tiles are simply high density fibre boards with either a thin laminated face, such as glass cloth or perforated vinyl, or a high-performance painted finish that maintains their surface porosity. Tiles commonly referred to as 'mineral fibre' are made of some combination of stone or glass wool, cellulose/recycled paper, clay, perlite, and starch, mixed together into a pulp and cured by heat. The tile is then finished with a water-based paint, laminated scrim and paint, or other decorative facing.

There continues to be a high level of innovation in the aesthetic development of ceiling tiles. With digital printing, tiles can now be produced with all sorts of colours, patterns, and visual effects including faux wood, leather, or metal finishes.

Not all ceiling tiles are acoustically absorptive; for example, gypsum, vinyl, or ceramic tiles are commonly used but provide no sound absorption. Each manufacturer should be able to provide an NRC rating of their tile, typically tested with a standard mounting of E-400 — suspended with a 400 mm air gap.
In addition to NRC ratings, ceiling tiles are often rated with a Ceiling Attenuation Class (CAC) rating. This tells you the tile's

Armstrong Fiberglass Tiles, left to right: Random Fissured Vinyl, Painted Nubby, Pebble, Shasta, Optima Health, and Lyra.

ability to block sound, which is desirable when you want to block noise from mechanical equipment in the plenum or stop speech transmission through the ceiling plenum between adjacent spaces with partial height walls. CAC is measured according to the standard ASTM E1414 and determines how sound transmits through the ceiling plenum between adjacent spaces. For this measurement, sound must pass through the tile in the source space, transmit through the plenum, and then pass through the tile in the receiver space. A CAC rating of less than 25 is considered poor, 25-35 is considered average, higher than 35 is considered high performance.

Generally speaking, fibreglass tiles have very high NRC ratings, between 0.7-1.0, but tend to have low CAC ratings (<25 CAC) because they are low-density and do not block sound well. Mineral fibre tiles tend to be higher density and will yield higher CAC ratings (20-40 CAC) but lower NRC ratings, between 0.5-0.8. Fibreglass tiles are sometimes provided with a metal foil backer to help provide modest CAC ratings. For some enhanced applications, fibreglass tiles may have a thin layer of gypsum board laminated to the back.

The Armstrong Health Zone Create line of tile allows for image printing and colours without reduction of sound absorption. The sample at left shows reference colour swatches; the sample at right is a Bamboo patterned print from the Wood Looks line.

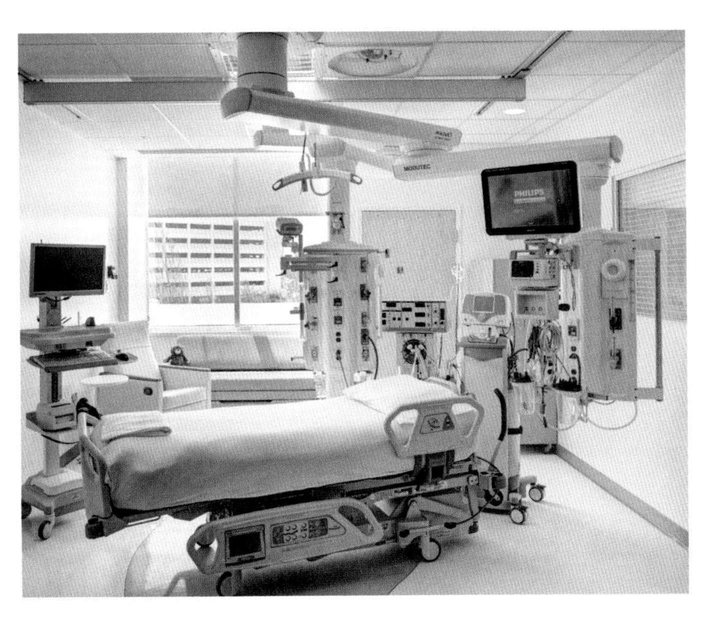

Nationwide Children's Hospital, Columbus, Ohio, designed by FKP (2012). The 12 storey hospital building was designed as a healing environment full of comforts and positive distractions for patients and visitors. A nature theme is found throughout. Colourful motifs of leaves, trees, and animals decorate each floor. Hallways feature educational exhibits and niches with fun animal facts. On the ground floor, a 'Magic Forest' and 'Aviary' designed by Ralph Appelbaum and Associates features a 3D surround soundscape by MorrowSound. The sounds of crickets, frogs, and birds change according to the time of day and even the time of year. Sound absorbing ceiling tiles are used in a variety of spaces throughout the hospital. Recent studies have shown the importance of acoustics in healthcare environments. Sound absorbing materials reduce stress and distractions, support sleep, increase patient privacy, improve speech intelligibility between staff, and lead to quicker recovery times.

ECOPHON

Ecophon, part of the Saint-Gobain group, is a line of acoustical ceiling and wall panels composed of high-density fibreglass. The panels are finished with a patented Autex surface technology, a specially formulated water-based painted surface that is optimized to allow sound absorption as well as provide light reflectance, light diffusion, gloss, and colour. The surface finish is smooth, visually resembling plaster or drywall. Panels can be installed in a ceiling grid or as independent units, wall panels, clouds, or baffles. Concealed grid options create pencil-thin seams between tiles for a more monolithic finish. Single panels can be fabricated in sizes up to 2400 mm x 1200 mm. Ecophon has also created a recycling technique that reuses dust from the production process, compressing it into their EcoDrain product, a lightweight aggregate which can be used as a drainage material at landfill sites in place of gravel.

40 mm thick Ecophon Solo Baffle.

Suspended circular clouds, made of Ecophon Solo, are installed in an open-plan office to provide a high degree of sound absorption and light reflectance.

HempFlax brand thermal insulation is made of 90 per cent hemp fibre reinforced with 10 per cent Bico or PLA (mais starch) fibre with soda added for fire protection. As hemp fibre does not contain protein, the material does not require chemical treatment against insects.

PLANT & ANIMAL FIBRES

Shifts towards sustainable, environmentally responsible, and recyclable building materials have piqued renewed interest in plant and animal fibres as alternatives to mineral and synthetic fibres for thermal insulation and acoustic absorption. A variety of plant and animal fibres were used in the development of some of the earliest acoustical materials nearly a century ago. Many kinds of fibrous materials can be used for sound absorption if they are optimised to provide the necessary flow resistance, which will relate to the fibre diameter, fibre orientation, density, and thickness.

The appeal of plant and animal fibres, when compared to manufactured fibres, is that they may require less energy for production and processing, they may have a positive environmental impact while they are grown, and they may be recycled without significant processing or energy consumption. On the other hand, plant and animal fibres do not tend to be inherently fire resistant and they may be susceptible to damage from moisture, fungus, insects, or vermin. In many cases, these concerns are addressed by incorporating additives to protect the fibres. Any additives or processing would need to be factored into the overall sustainability of the final product.

FLAX

There are two main varieties of flax; one grown for fibre production, the other for seeds. Flax grown for fibre is tall and straight, whereas the plants grown for seed production are short and branched. Harvesting can involve mechanised equipment such as combines, or can be done manually to maximise the fibre lengths. Flax fibres are extracted from the bast or skin of the stem and are stronger than cotton fibre but less elastic. Bundles of fibre have the appearance of blonde or 'flaxen' hair. The best fibre grades are used for linen fabrics such as damasks, lace, and sheeting. Coarser grades are used for manufacturing of twine and rope. Like cotton, flax fibre is a cellulose polymer, but its structure is more crystalline, making it stronger, crisper, and stiffer to handle, and more easily wrinkled. Flax fibres range in length up to 90 cm, and average 12 to 16 microns in diameter. They absorb and release water quickly; the primary reason why linen is comfortable to wear in hot weather.

HEMP

Hemp has been used for centuries to make rope, canvas sails, paper, and textiles. Two types of fibres are obtained from the plant's stalk — long (bast) fibres and short (core) fibres. The long, strong bast fibres are similar in length to soft wood fibres, roughly 70 per cent cellulose, and contain low levels of lignin which gives a plant rigidity and mechanical strength. The fibre diameter ranges between 16 to 50 microns. Hemp fibre conducts heat, dyes well, resists mildew, and has natural anti-bacterial properties. Shorter, woody core fibres, called tow, contain higher levels of lignin and are similar to hard wood fibres. When grown as a fibre crop, hemp grows easily to a height of 4 metres. Depending on the processing used to remove the fibre from the stem, the hemp may naturally be creamy white, brown, grey, black, or green.

BAMBOO

Bamboo fibre is a cellulose fibre extracted or fabricated from natural bamboo. Although bamboo can be processed in a traditional manner by retting, most bamboo fibre on the market is manufactured like rayon — a synthetic viscose made from bamboo cellulose. While bamboo can be grown and cultivated quickly without the use of pesticides or fertilisers, the process of manufacturing bamboo viscose can be a source of pollution.

JUTE

Jute is extracted from the bark of the white or tossa jute plants with fibre lengths of 1 to 4 metres and fibre diameters between 17 to 20 microns. It is low cost, yields four times more fibre than flax, and ranks second to cotton in terms of global production quantity. The product flourishes in tropical lowland areas with heavy rainfall and the main producers are India and Bangladesh. Jute has high insulating and anti-static properties, moderate moisture regain, and low thermal conductivity.

COIR

There are two types of coir fibre, both extracted from coconut husks: brown fibre, which is obtained from mature coconuts, and finer white fibre, which is extracted from immature green coconuts after soaking for up to 10 months. The food industry is the main consumer of coconuts, which primarily consumes the coconut meat. A fraction of coconut production is used for fibres, which are considered a by-product of the food industry. Coir fibres measure up to 35 cm in length with diameters of 12-25 microns. Coir has one of the highest concentrations of lignin among plant fibres, making it stronger but less flexible than cotton and unsuitable for dyeing. Coir has a good natural resistance to microbes and salt water.

MILKWEED

Milkweed fibres are obtained from the seeds of any of several milkweed plants of the genus Asclepias. Milkweed is a native perennial in North America. The seed pods produce a silky lightweight fuzz, called silk or floss. The lustrous, soft fibres are yellowish white in colour and remarkably similar to the hollow Kapok fibres, with fibre lengths around 2.7 cm, an outer diameter of 25 microns and a wall thickness of 2 microns. They have a natural buoyancy and were used as substitutes for kapok to make life jackets during World War II. Milkweed is hypo-allergenic, super-hydrophobic and highly oleophilic. The fibres are difficult to spin into a yarn but can be used as stuffing, in non-woven textiles, or blended with other fibres. The plant has long been perceived as a weed and its eradication has endangered the monarch butterfly whose caterpillars feed exclusively on milkweed. The Canadian company Encore3 produces milkweed primarily as an absorbent for oil spills. They are currently developing applications for milkweed in composites and thermo-formed parts for thermal and acoustic applications. The company strives to help bring back Monarch populations through the growth of milkweed. Pods are harvested in late September once the butterflies have migrated south. Milkweed is hearty and can be grown without pesticides.

KAPOK

Kapok fibre is the seed pod fluff inside the leathery pods of a rainforest tree called the Ceiba tree or the silk cotton tree. Kapok fibres are lustrous, yellowish brown, and made of a mix of lignin and cellulose. Each fibre is about 2.5 cm long with a hollow inner cavity. The outer fibre diameter is about 15-23 microns with a wall thickness of less than 1 micron. The fibres are naturally covered in a wax making them resistant to water but highly flammable. Kapok was used as stuffing in early life jackets prior to synthetic foams. The short fibre length makes it difficult to spin into yarn, so Kapok finds most applications as stuffing for mattresses, pillows, upholstery, and stuffed animals.

WOOL

Wool is a protein fibre, shorn from sheep usually once per year. After scouring to remove grease and dirt, wool is carded, combed, then spun into yarn. Fabrics made from wool have greater bulk than other textiles, providing better insulation. Wool is resilient, elastic, durable, and inherently resistant to flame and heat. Fibre diameters range from 16 microns in superfine merino wool to more than 40 microns in coarse wools. Two thirds of wool production is used for clothing; slightly less than one third is used for household textiles such as blankets, upholstery, carpets and rugs. Wool is also produced as an environmental alternative for thermal insulation in buildings.

CELLULOSE

Cellulose is the primary component of plant cell walls. For industrial applications, cellulose is obtained from wood pulp and cotton using the 'kraft process' to mechanically and chemically separate the cellulose from lignin. The resulting cellulose pulp is used to make paper and card stock and can be processed to create a variety of other materials including films and plastics. Cellulose also has thermal properties and is used for building insulation.

A company called Cellulose Material Solutions manufactures acoustical boards consisting of 65-75 per cent recycled content and resulting in no leftover scrap during the manufacturing process. The products themselves are also fully recyclable. The material is available in a variety of thicknesses and densities and its acoustic absorption performance is comparable to fibreglass board.

COTTON

Cotton fibres grow on the seeds of a variety of plants of the genus Gossypium. Of the four cotton species cultivated for fibre, the most important is G. hirsutum, which originated in Mexico and produces 90 per cent of the world's cotton. Cotton is almost pure cellulose, with fibre lengths ranging from 10 to 65 mm, and diameters from 11 to 22 microns. An estimated 60 per cent of cotton fibre is used as yarn and threads in a wide range of clothing. Acoustical cotton panels are readily available on the market, manufactured in various thicknesses, densities, and colours. These products typically use a high percentage of recycled content and are themselves recyclable. Additives such as borates are often used to provide flame resistance. Sound absorption performance is comparable to fibreglass or mineral wool products.

DENIM BATT

Recycled denim insulation is designed as an alternative to fibreglass and mineral wool for thermal insulation and sound absorption. The material does not itch or contain chemical irritants, and it is perforated so that it can be torn by hand for quick and easy installation. Denim batt uses up to 80 per cent post-consumer recycled fibres from denim clothing and it is itself recyclable. The fibres are treated with a boron-based fire retardant that also serves to impede the growth of fungus and mould. The primary manufacturer, BondedLogic, is developing many new product applications such as water heater jackets, pipe wrap, duct wrap, and multi-purpose rolls.

Opposite: Residence for A Briard, Culver City, California, designed by Sander Architects (2008). The home was built for a music lover who wanted the house to be a place where string quartets could play to a small audience. A prominent feature of the two-story great room is the denim batt ceiling, which is held in place by a thin wire mesh. The batt not only provides thermal insulation for the home but also sound absorption, controlling reverberation and creating a greater sense of acoustic intimacy.

Above: 3 mm thick industrial, medium density, pressed wool felt in grey, light grey, and white colours.

Opposite: Bloomx by FilzFelt is a modular screening system made from one repeated element designed by Italian graphic designer Chiara Debenedetti. The modules are 3 mm thick cut felt which interlock and connect to supporting headers above. The felt is available in up to 63 colours, allowing for infinite design possibilities. FilzFelt is a New York based company that specializes in acoustic materials, drapery and floor coverings, offering a wide array of colourful and ornate wool felt products as well as custom fabrication services.

FELT

Felt is a non-woven textile produced by matting, condensing, and pressing fibres to create a dense fabric of interlocked fibres. Felts can be made of natural fibres such as wool or synthetic fibres. Felt has been made and used by humans for centuries and it is one of the earliest materials to have been employed for acoustical absorption purposes. As with most fibrous materials, the absorptive performance increases with thickness and mounting away from the reflecting surface.

Felting is a manner of production that can have a wide variety of outputs for fashion, industry, craft, or furniture. Felts can be fabricated to be soft and pliable or tough and rigid. Felts can vary in terms of density, thickness, colour, fibre content, and many other factors. Felt can be made dense enough to be turned on a lathe or soft enough to be sewn. Felt cuts with a clean edge and does not ravel or fray. Wool felt may be cut by laser.

Wool felt in particular offers many attributes in terms of both design and ecology, lending itself to many acoustical applications in recent years. Wool is a sustainable and natural resource that is also biodegradable. Wool is naturally fire resistant and self-extinguishing.

Tribal DDB Amsterdam, The Netherlands, designed by i29 Interior Architects (2011). Tribal DDB is a digital marketing agency whose Amsterdam offices support a staff of approximately 80 people. The office design demanded a floor plan that was open and flexible, yet would allow for focused work and privacy. Balanced acoustics were critical in achieving this and the design team was attracted to felt not only as a sound absorber, but also for its durability, flame resistance, and sustainability. Felt is clad on a variety of surfaces, furniture, and even pendent lampshades throughout the offices. As an alternative to a suspended ceiling, the felt is strategically used to cover the underlying building structure which was scarred from past demolitions and remodels.

XSSS (Experimental Sound Saving System), W.M. Keck Lecture Hall Southern California Institute of Architecture (SCI-Arc), Los Angeles, California, designed by Hodgetts + Fung (2004). The sculptural ceiling treatment exploits the sound absorbing properties of 16 mm thick industrial felt to repurpose the concrete, industrial structure for use as a lecture hall and multi-purpose space. A unique pattern of slits was developed by the design team that enabled the felt to function as an elastic membrane, from which design constraints were extrapolated in the final design. Surface parameters were translated into 'sweeps' which oscillated laterally in order to complement the character of the surface, thus avoiding the visual interruption of an orthogonal module. The system was fabricated entirely with student labour. A PVC tubular armature system supports the felt, which is attached with vinyl upholstery buttons. Twelve downward-casting full-range speakers poke through the felt slits and provide distributed amplification during lectures. The space can be subdivided with movable partitions and is used for instruction, critiques, exhibitions, and presentations.

Evergreen moss panel in Apple Green colour by Freund.

MOSS

Natural moss can be processed and preserved to create a decorative and acoustically absorptive finish that is long lasting, maintenance free, and does not require artificial light, water, or fertiliser. The Berlin-based company Freund offers three different types of moss as suspended pendants or in a panelised format consisting of 50 mm thick moss pre-mounted to 10 mm thick MDF with concealed edges for simplified installation. Test reports indicate that 50 mm thick moss is an effective absorber at high frequencies above 500 Hz. The moss is available in two standard colours of Moss Green and Apple Green or may be dyed in a variety of custom colours upon request.

Weissglut, Neumarkt in der Oberpfalz, Germany, designed by Fellner Schreinerei (2015). Weissglut is a restaurant specialising in burgers and steaks using only organic, sustainable, and fresh ingredients. The interior is decorated with suspended moss balls (50 cm and 30 cm diameter) and a matching 17 m² moss wall, manufactured by Freund, that provides sound absorption and reflects the green and sustainable ethos of the restaurant.

GREEN WALLS

Green Walls, also known as living walls or vertical gardens, have surged in popularity over the past decade. Green Walls can be installed indoors or outdoors, free-standing or fixed to an existing structure, and can be constructed in various sizes with a variety of plant species. In addition to their aesthetic attributes, Green Walls can provide thermal insulation when installed on the exterior of a building, and may contribute to improved air quality.

The sound absorbing capabilities of Green Walls is an emerging research topic and there are many opportunities for further study. Preliminary research has indicated that the bulk of sound absorption in a living wall system is due to the substrate material that supports the plant life. Substrates may be composed of loose media such as soil within a compartment or shelf or soil held within a bag or non-woven fibre enclosure; other substrates may be composed of non-woven textiles, felts, coir fibre mats, or polyurethane foam. In order to absorb sound, the substrate should be porous: either a porous textile, membrane, or soil. Soils mixed with fibres or additives such as perlite, vermiculite, or pumice, will tend to be highly porous. Clays, on the other hand, tend to have low porosity and little absorptivity. A soil's sound absorption can change according to the amount of water saturating the soil. When saturated, the pores of the soil are filled with water and this will cause sound to reflect rather than absorb. Tests have shown that 50 mm thick porous soil can yield high absorption across the frequency range of hearing.

Plant life will cover the substrate and this coverage may reduce the substrate's sound absorption at high frequencies. The leaves and stems of the plants can scatter or diffuse sound, and they may also absorb through internal damping of sound-driven oscillations. These acoustic effects occur for frequencies above 500 Hz and are determined predominantly by the size, density, and angle of orientation of the leaves. The effects are consistent with the geometry of the plant, such that the larger the stems or leaves, the lower the frequency of sound that will be effected.

Opposite: Foundry Square III, San Francisco, California, architecture designed by Studios Architecture, living walls designed by David Brenner of Habitat Horticulture (2013). Foundry Square III is one of four buildings that comprise the urban campus of Foundry Square in San Francisco's financial district. The lobby features two adjoining living walls with 12,500 plants opposite two floor-to-ceiling glass walls.

MYCELIUM

Mycelium is the vegetative part of a fungus, consisting of a mass of interwoven, branching, filamentous hyphae. Fungal colonies composed of mycelium are found in and on soil. Mycleium can be grown around agricultural by-products, which provides both a food source and a base structure for the fungi. The mycelium fibres then create a binder throughout the agricultural material, resulting in a composite material. This process can produce flat materials or complex forms. Once the forming has taken place, the material is dried to stop growth and to prevent it from producing mushrooms or spores.

Ecovative is a pioneering bio-materials company and the leading developer of mycelium-based materials and applications that offer alternatives to non-renewable resource materials. Their products include MycoBoard panels for furniture and construction and MycoFoam packaging materials. Since mycelium products may be designed to provide sound absorption, Ecovative has begun developing acoustic tiles and panels for interiors. One of the primary features of mycelium-based acoustic boards is that the visual quality and texture can vary significantly depending upon the type of substrate material on which the mycelium is grown on. Ecovative offers a grow-it-yourself kit that allows designers to experiment with producing their own mycelium based materials.

EXPANDED CORK

Expanded cork was developed in the mid 1990s using parts of the cork bark that were not previously used. When the cork oak tree is stripped of its bark, generally only the trunk is stripped and the upper branches are left untouched. The cork bark from the upper branches cannot be used to make natural corks, as it is not thick or uniform enough and contains too much resin. Expanded cork makes use of these upper branches and can be stripped without chopping a tree down. This process is possible every 9 years for the total life of the tree, which is approximately 100 years.

Once this cork has been stripped from the tree, it is processed into granules which are placed in a mould and compressed. Super-heated steam is applied for 20-30 minutes at temperatures up to 370 ºC. The steam causes the granules to expand. Resins contained within these granules come to the surface and then serve as the adhesive that binds the granules together. The resulting product is a block of expanded or black cork which can then be cut into boards of varying thickness with densities ranging between 105-125 kg/m^3. Expanded cork is much darker than normal cork because of the resins and it also has a distinctive resinous odour that fades over time.

As a sound absorbing material, tests have indicated expanded cork primarily absorbs mid-to-high frequencies. Surface mounted panels have typically achieved NRC ratings between 0.4-0.6, depending upon the material thickness.

Post Office Ltd., London, UK, designed by HLW International (2015). The move to a new office building marked a cultural change in the workplace toward flexible and open spaces that would introduce new ways of conducting business. Heradesign wood wool was used in a concealed ceiling grid for touchdown spaces at the core of the office. The panels were colour-matched to the floor and wall colours, creating vibrant, dynamic blocks of colour that give each touchdown space a distinct identity and serve as landmarks to help guide people through the building.

WOOD WOOL

Wood wool products are manufactured by a number of companies around the globe, most notably Tectum in the United States, Troldtekt in Denmark, and Heradesign in Austria. The panels are made from wood fibres which are bound together with cement or a similar binder. Heradesign panels are bound together with magnesite, which is a less aggressive bonding agent than cement and keeps the panels more elastic.

While wood wool has been in use for many decades, it is experiencing a renaissance in recent years as designers such as Form Us With Love have discovered just how versatile this material is. It can be moulded into different forms, cut and shaped with standard woodworking tools, and painted or dyed. It is strong, impact resistant, and lightweight. Wood wool can be used as a support structure for finishes such as veneers or fabrics and can be combined with other materials to create composites. Tectum, for example, has a product line of roof decking material which combines wood wool adhered to rigid polystyrene insulation, providing both thermal and acoustic performance in one finished material.

Wood wool will typically achieve moderate sound absorption of approximately NRC 0.40–0.50 when A-mounted. For higher absorption performance, the material is often mounted with a porous backing layer.

The Tectum Finale panel is made of wood wool with a high-density mineral wool blanket integrated into the core of the panel. The composite design yields improved NRC ratings.

EchoPanel Geo Hex Tiles in assorted colours by Kirei/Woven Image. The tiles are provided with a peel-and-stick option for quick installation.

PET

PET, polyethylene terephthalate, is a thermoplastic resin and the most common type of polyester. In addition to its use in polyester textiles, PET can be extruded or moulded into plastic bottles and packaging containers for food, beverages, and other consumer goods.

PET is completely recyclable and PET products are identified with a number 1 inside the triangular recycling logo that is moulded into most plastic packaging. Recycled PET can be reformed into packaging or components for products and automotive parts. It may be recycled into fibres for carpets, fibrefill for upholstery or clothing, as well as acoustically absorptive materials.

A number of acoustical products have appeared in recent years using PET for panels, tiles, fabrics, dividers, and furniture. The materials are usually made from some degree of recycled material and are themselves fully recyclable at the end of their use. PET materials can be manufactured to various thicknesses and densities and blended with other fibres. PET can be thermally bonded and formed without the use of adhesives or binders. Some product lines use a dense layer of PET for structural stability bonded to a lighter density that provides the bulk of sound absorption. Panels may have a decorative polyester fabric bonded to the face of a PET core. Some products' densities allow the material to double as a bulletin board.

PET is non-toxic, non-allergenic, and contains no irritants. The material feels somewhat like dense felt and it does not require protective coverings or facings like fibreglass or mineral wool.

EchoPanel Wave Tile by Kirei/
Woven Image.

ide Group, a leading Australian product development and design firm specialising in consumer, technology, and medical device markets, designed the geometric and functional acoustic wall installation for their Sydney based headquarters. Six basic shapes were cut from 50 mm thick PET Ecoustic Panels by Unika Vaev/Instyle. The shapes were then pieced together in different formations to create a unique installation for each room. The material is optimally placed on the wall at the height of a seated and standing person.

Impressions is a wall-mounted acoustic tile system where six classic KnollTextiles patterns are 'impressed' onto a high performance, ultra-thin 100 per cent polyester panel, available in 16 standard colours. Impressions is ordered as a complete kit with Z-clips and 3.5 cm thick rails for mounting. Each tile measures 30.5 cm x 30.5 cm x 10 mm and when installed on the rail system achieves NRC 0.55.

Opposite Above: Snowsound BAFFLE in standard sizes and colours.

Opposite Below: Snowsound PLI divider screen.

SNOWSOUND

Snowsound acoustic panels were designed by architects and designers with the goal of creating a product that is as stylish as it is functional. Manufactured in Italy, the panels are 100 per cent polyester with a bonded Trevira CS polyester fabric surface that is visually soft, yet highly durable and resistant to impacts or tearing. The patented method of fabrication and bonding, using different core densities of material, creates a lightweight, seamless panel that is very strong and does not require any additional framework for strength or stability. Since the panels are made of a single material, polyester, they are entirely recyclable at the end of their life cycle.

The various methods of panel attachment also exhibit a high level of design consideration that allows for flexbility, variety, and modularity. The FLAP system, for example, uses a metal stand-off with a pivoting head that allows the panels to be tilted or rotated in a variety of patterns. The PLI sound panels can be transformed into stand-alone room dividers by using a simple elastic band system.

FOAMS

Foams are formed by trapping pockets of gas within a liquid or solid. Foams can be categorized as either open-cell or closed-cell structures. With closed-cell foams, the gas forms discrete pockets that are fully surrounded by material. In open-cell foams, the gas pockets create an interconnected network of cavities. A kitchen sponge is an example of open-cell foam, where water can displace air and penetrate the entire structure. A yoga mat is an example of closed-cell foam, the gas pockets are self-contained which may create a cushion and will not allow water to permeate the interior structure.

Open-cell foams can be excellent sound absorbers due to their porosity. Closed-cell foams are not typically good absorbers because they do not allow sound to penetrate into the material. However, in some cases closed-cells can be perforated or burst open after they are manufactured, creating interconnected cells that makes them function more like open-cell foams.

Acoustitch, designed by RCKa, St. James's, London (2013).
The installation in the foyer of a building in Waterloo Place is crafted from 780 brightly-coloured foam wedges, inspired by anechoic test chambers. The wedges attenuate the reverberation time within the space and provide an infrastructure for a changing geometric pattern which references the area's rich history of tailoring. As visitors to the building traverse the reception area, the appearance of the piece subtly shifts revealing two alternating colourful weave patterns, depending on the point from which it is viewed.

FLEXIBLE POLYURETHANE FOAM

Flexible Polyurethane Foam is used in a wide variety of consumer products that impact our daily lives such as bedding, furniture, packaging, textiles, automotive interiors, filters, and shoes.

Polyurethanes are formed by the reaction of two chemicals — polyol, a type of complex alcohol, and diisocyanate, a petroleum by-product that reacts strongly with the polyol alcohol. The result is a polymer, or plastic, known as urethane.

When the polyurethane is in a hot liquid state, it is mixed with carbon dioxide while being sprayed into a thin layer. The carbon dioxide gas forms many tiny bubbles within the liquid. As the polyurethane is heated to dry, the gas bubbles expand and burst, resulting in an open-cell porous structure.

Polyurethane foam is relatively inexpensive and is commonly used in home recording studios. Polyurethane foam is not often used in public spaces because one of its primary drawbacks is its flammability and smoke development ratings. The foam can be treated for fire-resistance but still may not meet the stringent flammability requirements of some jurisdictions.

REBONDED POLYURETHANE FOAM

Rebonded foam is produced from chopped, flexible polyurethane recycled scrap that is held together with a binder. This type of foam is commonly used for carpet underlayment padding as well as packing material, furniture upholstery, and footwear. Rebonded polyurethane foam is flexible, workable, and sturdy. Other acoustic applications include automobile interiors and inside machinery housing to reduce noise from equipment such as compressors, engines, and vacuum cleaners.

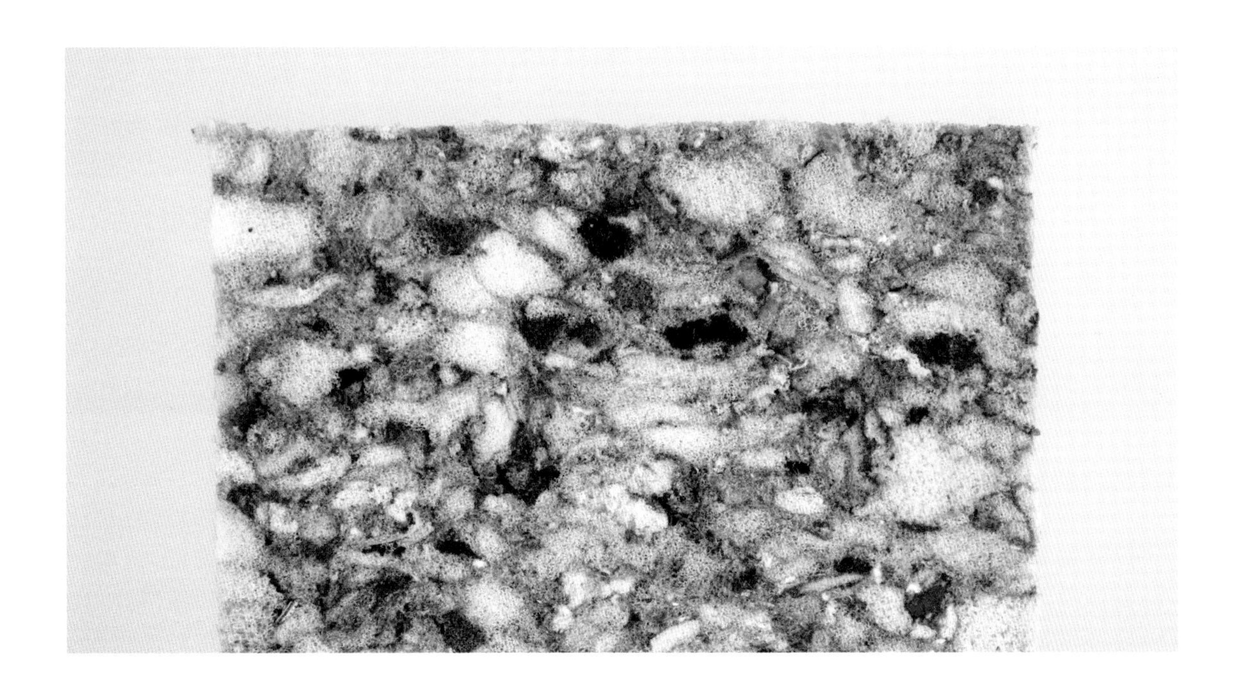

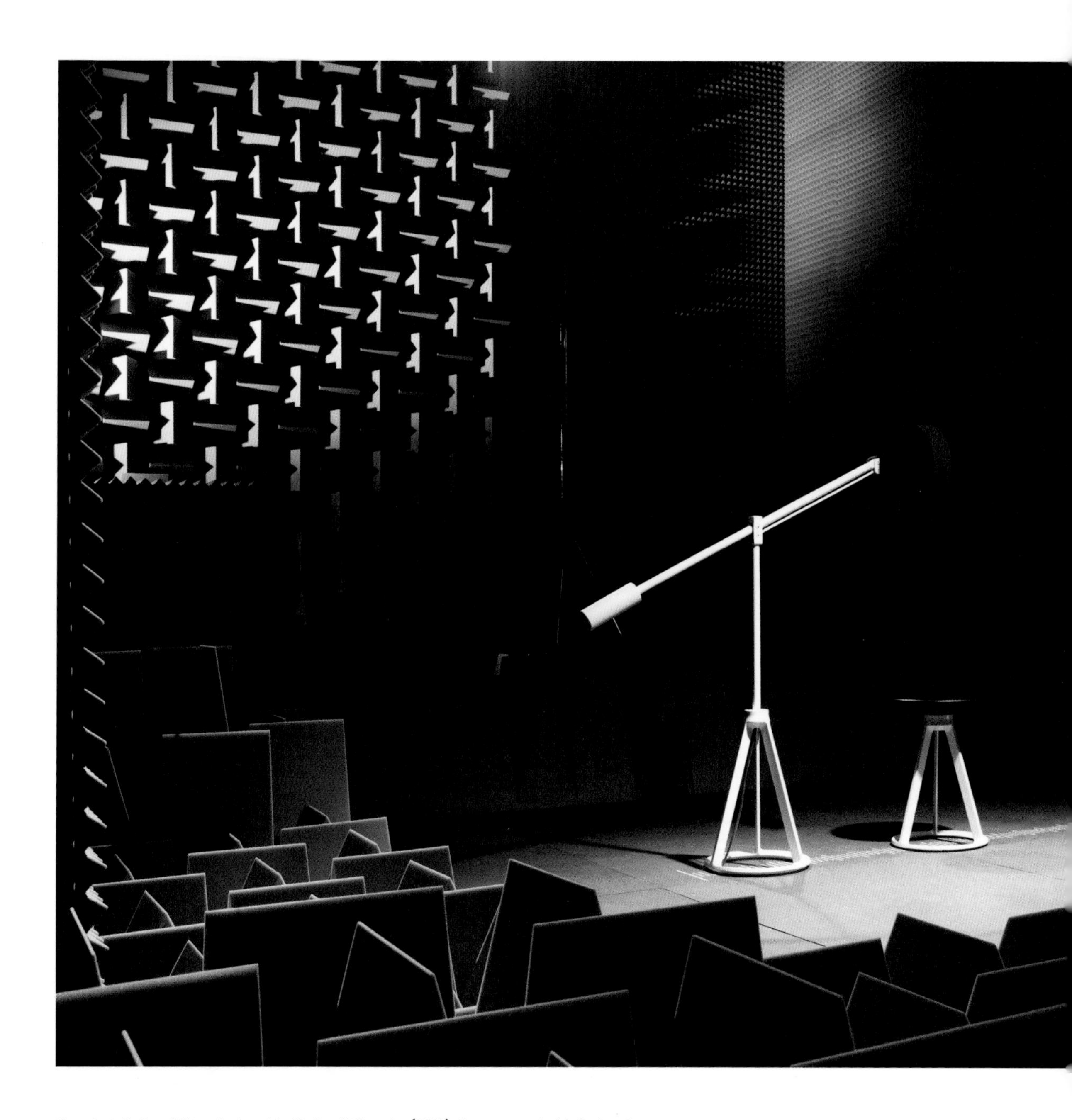

Sony Installation, Milan, designed by Barber & Osgerby (2010). Sony teamed with Barber & Osgerby to design their first stand-alone exhibition at the 2010 Milan Furniture Fair, which showcased the latest Sony conceptual developments integrating electronics within furniture and interior spaces. The exhibition was titled 'Contemplating Monolithic Design,' exploring the essential elements necessary for products to create a sense of presence in a space. The exhibition space was subdivided into five areas, organising the experiments into archetypes: icon, insight, intimate, integrate, and install. The darkened anechoic space invited visitors to enter a laboratory-type chamber with a heightened sense of hearing and seeing. Near-field speakers were integrated into seat headrests to create intimate entertainment spaces; a TV cabinet was built from a 3D printer with a porosity that allowed sound to transmit from speakers concealed inside the cabinet; and a series of lights and sculptural objects utilised transducers to transform them into sources of sound.

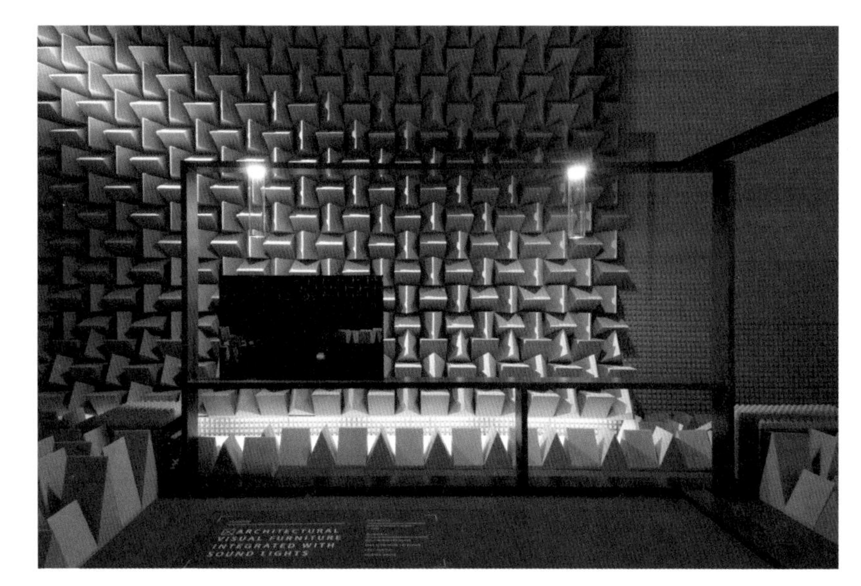
ARCHITECTURAL
VISUAL FURNITURE
INTEGRATED WITH
SOUND LIGHTS

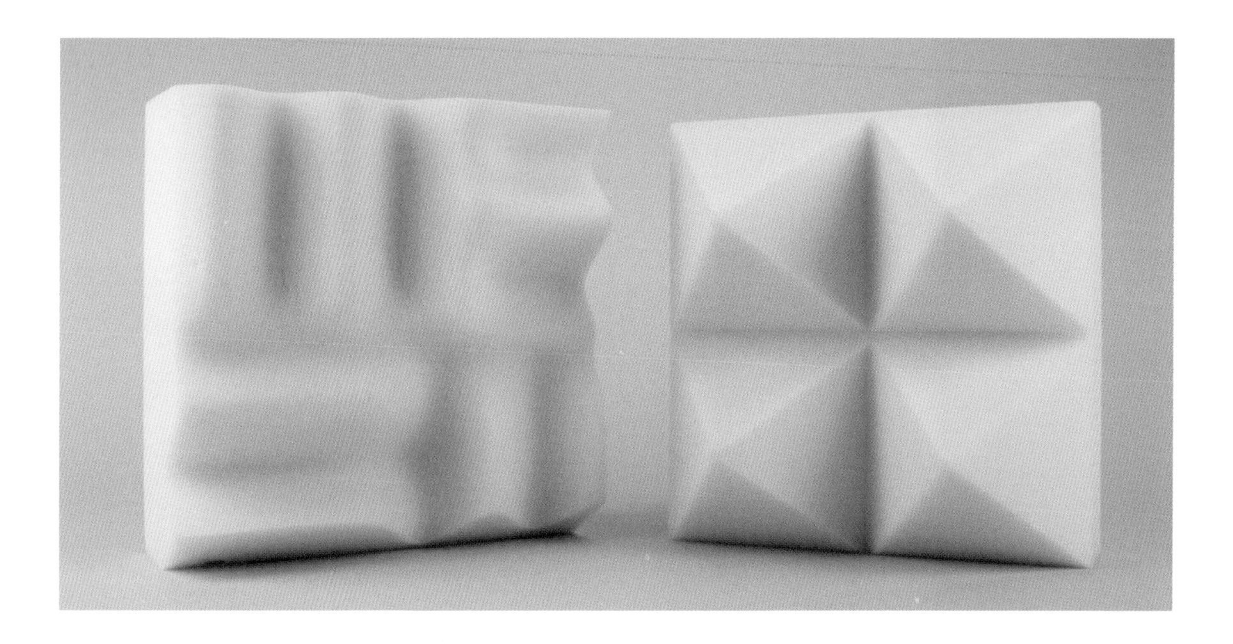

Sonex and Pyramid patterned white melamine foam by Pinta.

MELAMINE FOAM

Melamine foam is an open-celled foam composed of a formaldehyde-melamine-sodium bisulfite copolymer that is used for thermal insulation, sound absorption, and everyday products such as the Mr. Clean Magic Eraser by Procter & Gamble.

The high nitrogen content of the resin used to produce the foam results in an inherently flame-resistant material without the need for additional flame retardants. The material can be formed or cut into 3D shapes, as well as flocked, printed, and embossed. The foam may also be laminated to other materials such as felt, fabric, metal, or plastic.

One of the largest producers of melamine foam is the German company BASF. Acoustical melamine foam is primarily sold internationally under Pinta Acoustics which provides the foam in standard colours of grey and white and in standard formats such as flat sheets, wedges, pyramids, and cylinders.

Melamine foam has a micro-abrasive surface texture, which is why it works well as the Magic Eraser, but this makes it difficult to maintain when installed as an architectural finish. For this reason, the melamine foam is offered with a factory-applied high-performance coating that protects the material and allows it to be wiped clean. Without the coating, raw melamine foam may be cleaned by lightly vacuuming the surface. The high-performance coating does not significantly alter the sound absorption performance, and it is available in a variety of standard and custom colours.

White melamine foam with Sonex pattern and a dark grey high-performance surface coating by Pinta.

Flat melamine foam in standard colours of grey and white.

Alchemist Boutique (Ground Level), Miami, Florida, designed by Rene Gonzalez Architect (2009). Nestled on the ground floor of the Herzog and de Meuron-designed 1111 Lincoln Road parking garage, this concept store is designed as an enveloping and protective cocoon-like space that allows patrons to disconnect from the motion, sound, and heat of the outdoors. Melamine foam envelops the ceilings and walls, acting as a buffer for the surrounding structure and providing a soft, tactile quality. A symmetric and rhythmic pattern of the saw-tooth patterned foam walls is accentuated by hidden LED light bands that produce a choreographed and changing colour composition that appears and disappears behind the foam.

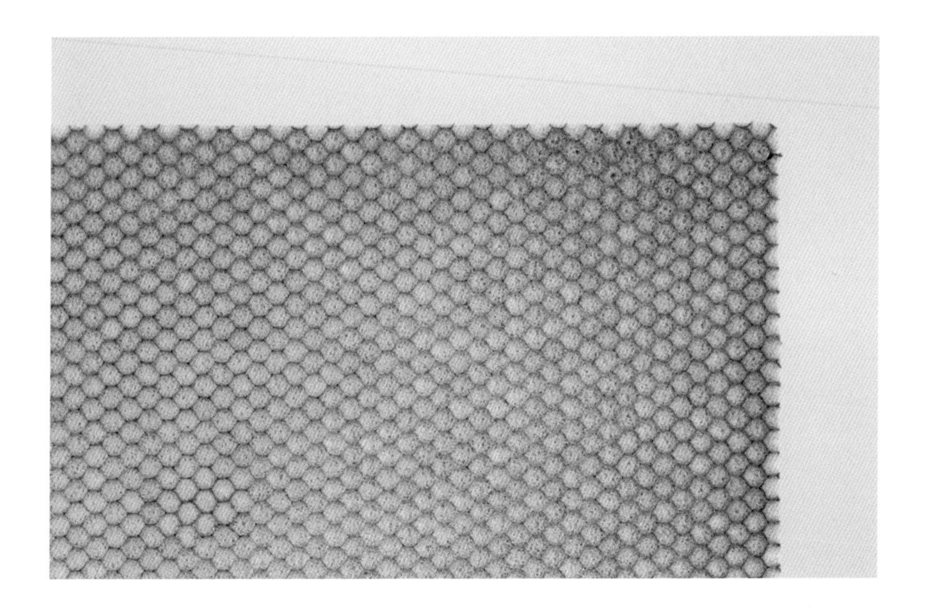

FOAM-FILLED HONEYCOMB

Foam-filled honeycomb combines the lightweight sound absorption performance of foam with the structural stability of a honeycomb core. The resulting composite material can provide improved thermal insulation, enhanced sound absorption, and higher resistance to impacts than either material could provide independently.

One method for filling honeycomb with foam is simply to press the foam into the empty cells. This process only accommodates lower density foams, which are friable and may generate dust and wasted material. The dust may also interfere with the bonding of the core to facing materials.

Another method is to pour un-foamed resin into each honeycomb cell and then allow it to foam. With this process, it can be difficult for the foam to achieve a uniform density - the foam may be more dense at the bottom of the cell and less dense at the top.

M.C. Gill Corporation has been manufacturing honeycomb core products since the 1980s, and has optimized foam-filled honeycomb production with the recently introduced product GillFISTS. During the manufacturing process, a proprietary coating is first applied to the honeycomb core. Once heat is applied, the coating generates foam within each cell. Heat application may occur before or after the outer finish facings have been applied to the honeycomb. The process results in a uniform foam density without dust to interfere with the bonding of facing materials. The product was originally engineered for aircraft use and is lightweight with a high fire resistance.

POLYIMIDE FOAM

Polyimide foam was first designed by NASA to create a relatively low outgassing, thermal insulating, and sound absorbing material. Boyd Corporation is the primary manufacturer of polyimide foam, which it markets under the name Solimide.

Polyimide foam is fire resistant, which makes it desirable for applications where other types of foam insulations may not be permitted. The foam emits virtually no smoke or incapacitating toxic by-products when exposed to open flame. Polymide foam may be used to line HVAC ductwork for projects where fibrous materials such as fibreglass are not permitted.

Polyimide foam is also extremely lightweight, with a density of roughly 6 kg/m^3. Polymide is a desirable option for marine, aerospace, and other forms of transportation since it is many times lighter than mineral wool or fibreglass.

POROUS EXPANDED POLYPROPYLENE BEAD FOAM

Porous expanded polypropylene bead foam, PEPP, is actually made of closed cell foam cylinders. The air pockets or voids created when the cylinders are formed together create the porosity necessary for sound absorption to occur.

PEPP is non-fibrous, structurally rigid, water resistant, non-abrasive, impact resistant, lightweight, and is resistant to fire, water, and bacteria. PEPP is also resistant to oils, grease, and most chemicals and is washable with a power washer. The durability of this material offers excellent applications in mechanical rooms, natatoriums, and equipment enclosures. Due to its light weight, PEPP has been used in car interiors to reduce noise levels. The panels are also tackable and may serve a dual function in conference rooms and classrooms.

PEPP is available in white or black and in standard thicknesses of 25 and 50 mm. The material can be fabricated into different shapes and with surface patterns. When surface mounted, a 25 mm thick panel yields an NRC rating of 0.55. The use of an additional acoustical backing material such as fibreglass or cotton board can yield even higher acoustical ratings.

WHISPER FOAM

Whisper foam is a closed-cell polyethylene foam whose cells are opened during the manufacturing process, allowing for sound absorption to occur. The foam is manufactured 25 mm thick, which is relatively bendable and can conform to curved shapes. Layers of the material may be laminated together to provide increased dimensional stability and compressive strength for panelised applications. Whisper foam is hydrophobic and ideal for marine and industrial applications as well as equipment enclosures. As with many foams, it may be cut and worked into custom shapes. The material is available in Natural, White, Black, and Grey colours. A 50 mm thick panel, when surface mounted, will yield an NRC rating of 1.0.

EPDM SPONGE RUBBER FOAM

EPDM (Ethylene Propylene Diene Monomer) sponge rubber is a closed or semi-closed cell foam. Despite the closed cell structure, it can exhibit moderate sound absorption, and it is commonly used in equipment enclosures or for lining HVAC ductwork when fibrous or open-celled materials are not allowed. Since it is closed cell, it will not accumulate dust or absorb water like most porous materials. One of the leading products is AP Armaflex by Armacell, which is formaldehyde free, low-VOC, and features an anti-microbial finish.

METAL & CERAMIC FOAMS

Metal and Ceramic foams are relatively new materials which can be designed for sound absorption but also offer a variety of engineering applications such as filtration, heat exchangers, lightweight structures, thermal insulation, and medical implants. These types of foams are very stiff, lightweight, resistant to very high temperatures and do not absorb moisture. Many of the materials can be recycled at the end of their life cycle.

Open-celled foams are often produced using a polyurethane foam as a skeleton structure that is replicated by the ceramic or metal material. For ceramics, the polyurethane foam is impregnated with a slurry that is then fired in a kiln, leaving only the ceramic structure behind. Ceramic foam can be made from a variety of materials such as silicon-based elements, aluminium-oxide, zirconium, titanium, and boron. While aluminium is one of the most popular metal foams, there is a variety available such as copper, nickel, zinc, steel, and titanium.

Closed-cell foams can also be produced, but they are not effective sound absorbers. However, closed-cell foams can have their cells broken open through rolling, drilling, or water jet to create an interconnected network of cells and thereby create sound absorption.

Other methods for producing these foams include injecting gas to create bubbles in a liquid melt, adding foaming agents which cause gas bubbles to generate in a liquid melt, or powder-based sintering. Many of these foamed materials can be expensive to produce and tend to be used on relatively small-scale applications.

Samples of foams manufactured by Ultramet, from left to right: Silicon Carbide with 20 pores per inch; Vitreous Carbon with 30 pores per inch; Silicon Carbide with 80 pores per inch.

ALUMINIUM FOAM

Stabilised Aluminium Foam (SAF), manufactured by Cymat Technologies and distributed under the name Alusion, is made by injecting air into a molten aluminium melt containing a fine dispersion of ceramic particles. These particles stabilise the bubbles formed by the air, much like dry cocoa powder stabilizes bubbles when it is added to milk. The manufacturing process uses up to 50 per cent recycled content and the resulting product is 100 per cent recyclable. Standard sheet sizes are 1220 mm x 2440 mm in thicknesses of 12.7 mm, 25.4 mm and 43.2 mm.

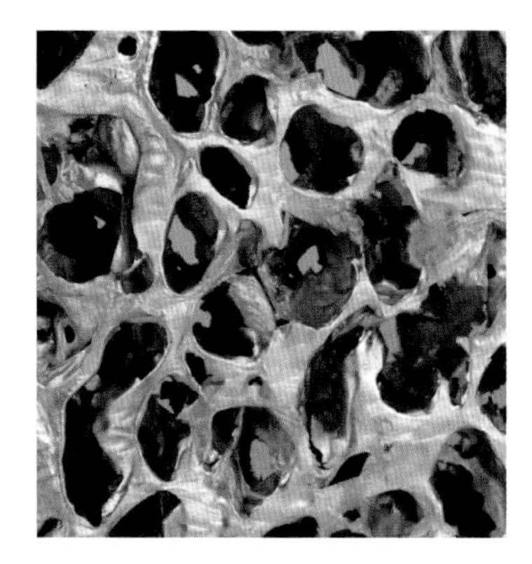

The manufacturer can custom-cast the material in a range of densities (0.11 g/cm³ to 0.55 g/cm³), thicknesses (12.7 mm to 88.9 mm), and even cell sizes according to specific applications. Porosity can be tuned between 80 to 96 per cent porous. The standard small, medium, or large cells typically have a thin-walled skin, which can be removed on one or both sides of the material with a special blasting process.

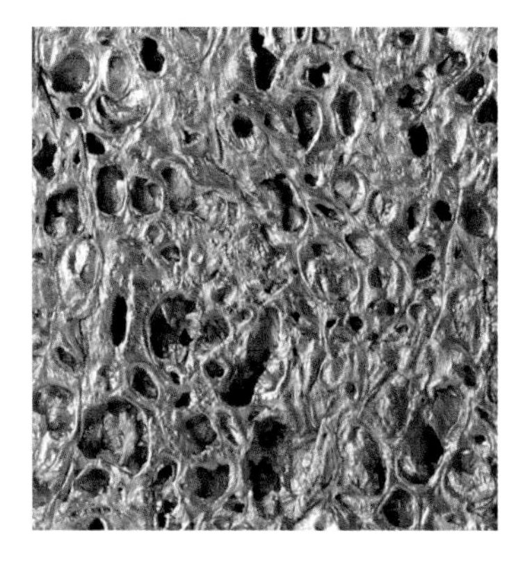

SAF is surprisingly customisable and workable. Powder coating finishes are available in limitless colours. Special powder coats can provide additional resistance to salt and chemicals. The material can be water-jet cut with shapes, patterns, logos, and letters. Panels can be bent using heat on the back surface or with a three roller bending machine. The material can be cut on-site using a large-toothed, carbide or diamond tipped circular saw. A band saw can be used for trim work and short cuts. Water or other liquid cooling lubricants must be avoided as they will cause discolouration.

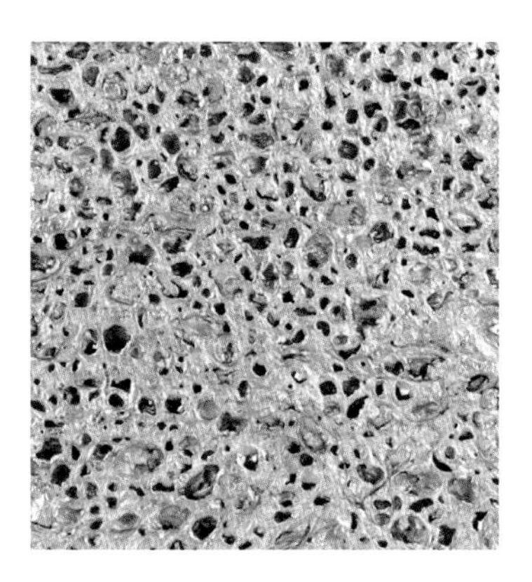

Acoustical test reports of standard SAF panels suggest absorption occurs primarily at high frequencies from 1 kHz and above, depending upon the mounting condition. Absorption performance can vary depending upon the porosity or cell size. Very high levels of broadband absorption can be achieved when open cell SAF is mounted with a backing layer of fibreglass, mineral wool, or similar porous material.

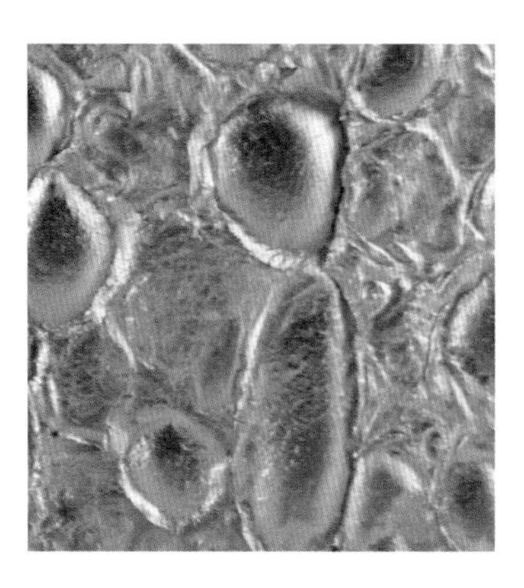

Stabilised Aluminium Foam samples from Alusion by Cymat Technologies, from top to bottom: Large Cell Open, Mid Cell Open, Small Cell Open, Large Cell Natural.

Fondazione Prada, Milan, designed by OMA (2015). The new Milan venue of Fondazione Prada, conceived by the architecture firm OMA — led by Rem Koolhaas — expands the repertoire of spatial typologies in which art can be exhibited and shared with the public. Characterised by an articulated architectural configuration which combines seven existing buildings with three new structures (Podium, Cinema and Torre), the new venue is the result of the transformation of a distillery dating back to the 1910s. Historic and new architecture, although separate, confront each other in a state of permanent interaction. The Podium building, designed to host temporary exhibitions, is clad on both the exterior facade and interior walls and ceilings with architectural aluminium foam. The Podium's aluminium texture complements the adjoining 'Haunted House', covered in gold leaf. Fondazione Prada, founded in 1993, is an institution dedicated to art and culture.

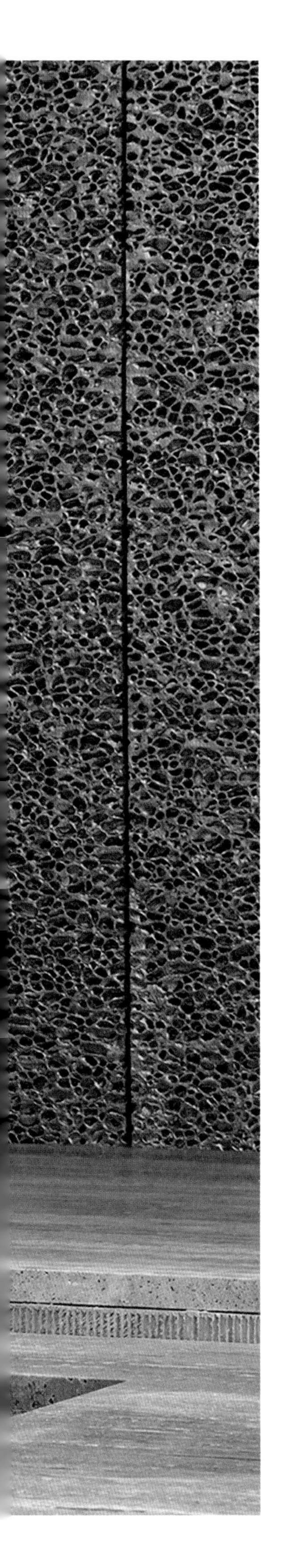

WOOD FOAM

Fraunhofer Institute for Wood Research has developed a new wood-based foam that is lightweight, porous, and composed nearly entirely of hardwood, softwood, or other lignocelluloses. The product is not yet available on an industrial scale, but will likely be an important construction material in years to come.

The material is produced from a very finely milled wood fibre suspension. This suspension is either mixed with a foaming agent, or the porous structure is created through whipping or strongly stringing. Finally, the foam hardens in an oven. Cohesion is achieved through the bonding powers of the wood itself, which are activated during the production process. The foam does not contain any synthetic bonding agents or adhesives.

The result of the processing is a lightweight base material with a porous, open-cell structure and low bulk density. The foams can be manufactured in a variety of densities ranging from 40 kg/m^3 to 200 kg/m^3. The foam is hard and stable and can be cut and processed just like any other wood material.

From an ecological and economic perspective, wood foams present a variety of potential applications. Preliminary sound absorption measurements in an impedance tube have shown that wood foam can function as an excellent sound absorber due to the porous, open-cell structure. The absorption performance can vary depending upon the density or porosity of the foam produced. It is likely that wood foam will find a variety of acoustic applications in the near future.

Lumira Aerogel particles and a Silica Aerogel cube measuring 2.5 cm x 2.5 cm x 1.0 cm, 0.095 g/cm^3 density.

AEROGEL

Aerogels are created by removing the liquid content of a gel through a supercritical drying process, which results in a highly porous, monolithic, lightweight and translucent material that is mostly filled with air. Several materials have been used to produce aerogels but silica is presently the most common. Since aerogels can be made from many kinds of materials, the term aerogel refers to a category of materials with a specific structure.

To date, aerogels have been rather expensive to produce and their applications have been limited to high-tech aerospace projects. As the production costs lower, aerogels are beginning to see their integration into more applications and composite materials such as layered into laminated glazing for thermal insulation or incorporated into blankets for thermal insulation and acoustic absorption.

It is possible to vary the porosity of aerogels in the production process and, therefore, their performance characteristics can be designed for specific applications. Aerogels can also be used in a granular form, in which case their mechanical behaviour is somewhere between a gel and a granular material. Aerogels hold a lot of promise for their strength, low thermal conductivity, and micro-porosity, and it is anticipated that aerogels will appear in many acoustical applications in the future.

POAL

Poal is a thin sheet material composed of non-woven aluminium fibres sandwiched between a metal mesh. The material is workable and can be mechanically cut, laser cut, drilled, and formed to different shapes.

Poal is non-corrosive, non-flammable, and can be used indoors as well as outdoors. It is composed partially of recycled aluminium and can be entirely recycled at the end of its life cycle. The material does not hold moisture and will absorb sound even when wet.

The product is available in a variety of custom factory-applied colours and specialised coatings including PTFE, fluoric resin, and polyester resin paints. Panels are available in slightly differing densities and thicknesses, with a thicker expanded metal mesh option, and an option for an inner aluminium foil layer. When the material is installed in locations susceptible to high-impacts, it may be mounted with an additional perforated MDF or sheet metal layer to provide added stability.

The sound absorption performance of Poal is dependent upon providing an air gap behind the material. The peak absorption will change according to the depth of the air cavity. For the best broadband absorption performance, Poal may be backed with an additional absorptive material such as mineral wool, fibreglass, or foam.

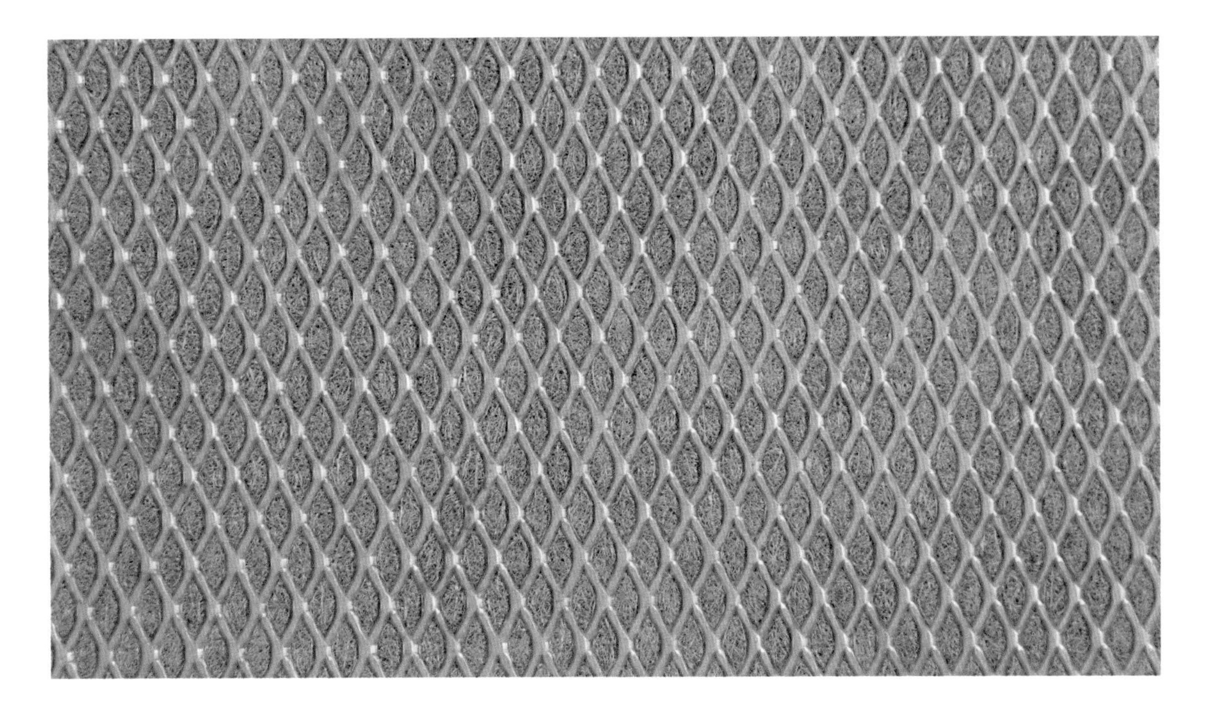

Above: Samples of Calme/Almute in Corrugated, Embossed, Intaglio, and Smooth Finishes.

Below: A detail of the Intaglio style surface finish.

CALME/ALMUTE

Sintered aluminium panels are marketed under the name 'Almute' in the United States and 'Calme' in Europe and Asia. The panels are produced through a sintering process of aluminium grains mixed with alloyed metal powders, which is spread over a carbon tray and heated in a furnace at near-melting temperatures. As the panels cool, large pores between the aluminium particles remain, resulting in a porous, acoustically absorptive material.

Following the sintering process, the panels are spray-painted to provide a smooth, uniform surface. The paint layer is limited to a thickness of 20 microns, which does not significantly influence the sound absorption performance. The standard paint colour is grey, but custom colours may be applied.

Panels are available in two standard sizes, which are limited by the size of the kilns in which the material is produced. The material can be cut to any specified smaller size and is offered in three types: (A) for indoor architectural applications; (B) for indoor applications but with the ability to be bent or provided with an embossed or corrugated surface; (C) for use outdoors or in corrosive or wet environments.

Calme/Almute should be mounted with an air gap to provide sound absorption. Absorption increases according to the depth of the backing cavity. Calme/Almute will achieve the highest broadband absorption performance when a porous material is layered in the backing cavity.

Tribunal of Lorient, Lorient, France, designed by Murisserie (2014). Courtrooms are often designed to support a high level of speech intelligibility. In the Tribunal of Lorient, Calme with a Stucco finish has been used for sound absorption on nearly every surface - the walls, ceiling, and furniture partitions.

QUIETSTONE

Quietstone FR30 is made from stone aggregate that has been bonded together with a resin. The size and shape of the aggregate, when bonded together, results in irregular tiny air cavities, creating an open porous structure with a high degree of tortuosity. The resin has been carefully crafted so that it gives the material high strength but does not seal the porous cavities.

The panels are extremely durable and particularly suited for use in outdoor and harsh environments. The material has been used for lining highway tunnels and was designed to meet stringent European tunnel safety requirements such as high impact resistance, moisture and frost resistance, and exposure to temperatures of up to 1150 °C for a duration of two hours.

Optimal sound absorption performance is achieved when the material is installed with an air gap and an absorptive backing material such as mineral wool or fibreglass. Panels can be pigmented in a variety of shades from off-white to true black.

REAPOR

Reapor is made entirely from small aerated granules of recycled glass which are fused together through a patented high-temperature sintering process. The granules themselves are porous and once formed into a panel, additional pores are provided between the granules, resulting in excellent broadband sound absorption performance. A 50 mm thick surface mounted panel yields an NRC rating of 0.90.

The panels can be cut, drilled, and machined using common woodworking tools and may be painted or tinted using water-based paint. Anti-graffiti and rendered finishes are also available. Panels are commonly installed using adhesives or mechanical fasteners. The material is homogeneous, lightweight, fibre-free, binder-free, non-combustible, UV-resistant, 100 per cent recyclable, and impact-resistant.

Reapor may be used outdoors and exposed to sun and rain. Wet panels will drain freely and dry in the sun; however, this can result in efflorescence where crystalline salts are deposited on the surface of the panel. Efflorescence will not affect acoustic performance and may be removed using commercial efflorescence cleaners.

VERMICULITE

Vermiculite is a naturally-occurring mineral that resembles mica. The crystal structure of vermiculite contains water molecules. In a process called exfoliation, vermiculite is heated under high heat and the water converts to steam, expanding the material up to 20 times its size. The resulting exfoliated material is porous, lightweight, chemically inert, fire resistant, with low density and low thermal conductivity. Vermiculite ranges in size from very fine particles to large, coarse pieces nearly 25 mm in length.

Expanded vermiculite is used in numerous products, including insulation for attics and walls or mixed into plasters and concrete materials, often combined with other lightweight aggregates such as perlite. Certain plasters containing vermiculite combined with binders like gypsum or Portland cement can provide both fire protection and sound absorption. The sound absorbing properties of exfoliated vermiculite depend on the shape and size of the vermiculite grains and the total thickness of the material.

SPRAY-APPLIED FIREPROOFING

Certain types of construction classifications require fire protection, which can be active (fire sprinklers) or passive (insulating structural elements from the heat of a fire). For passive protection, it is common to use spray-applied fire resistive materials on structural steel elements, columns, and metal decks to achieve the code required fire ratings. These spray-applied materials may be fibrous, gypsum-based, or cementitious plasters.

Many cementitious plasters are mixed with inorganic lightweight aggregates such as vermiculite or perlite to reduce weight and density. Some of these fireproofing sprays are porous and can additionally provide acoustical absorption and thermal insulation. Cafco Blaze-Shield, for example, is an inorganic, Portland cement based, medium density spray-applied fire resistive product that is formulated to withstand indirect weather exposure conditions and limited physical impacts. A 13 mm thick application of this material can achieve NRC ratings of 0.75-0.85. When such fireproofing material is left exposed to a space, it can be strategically and economically used to absorb sound. The resistance to weather and impacts makes it particularly useful for applications in parking garages, mechanical rooms, and similarly harsh environments.

K-13 samples in the standard colours of White, Light Grey, Beige, and Cocoa.

SPRAY-APPLIED CELLULOSE

Spray-applied cellulose consists of prepared cellulose fibres, composed of 80 per cent post-consumer materials which are chemically treated in accordance with a quality-controlled process. Licensed applicators use manufacturer-approved machines and nozzles to combine the fibres with a patented binder during the application process. The cellulose material can be applied to virtually any properly prepared common construction surface such as wood, steel, plastic, concrete, or glass and can conform to complex surfaces such as barrel vaults or corrugated metal decks. Spray-applied cellulose has been around for many decades and is commonly available under the names K-13 or SonaSpray by the International Cellulose Corporation. Spray-applied cellulose is cost-competitive for large applications. The sound absorption performance increases with the applied thickness and may depend in part on the substrate to which the material is applied. A 25 mm thick application can achieve NRC ratings of 0.9.

K-13 can be applied up to 125 mm thick in one application without mechanical supports. K-13 may also double as a thermal insulator, capable of providing R-values up to R-19. For areas such as indoor pools and ice arenas, the proper combination of K-13 and ventilation prevents condensation on metal, concrete, and other surfaces. It is available in six standard colours and a variety of custom colours.

K-13 is available with a special surface finish coating called 'Protek-13' for areas where physical damage or contamination from dust, dirt, or grease is a concern. Protek-13 is a spray-applied, water-based vinyl acrylic emulsion containing interlacing fibres that form a protective coating without significantly degrading the acoustic absorption performance. The combination of K-13 with a Protek-13 finish is commonly used in parking garages and similar harsh environments.

SonaSpray is similar to K-13 but provides a smoother finish texture. It is often used in areas with lower ceiling heights or in areas of high traffic and visibility. The adhesive used to apply SonaSpray provides a durable and resilient bond that is resistant to impact and abrasion and dose not pose the risk of cracking or spalling which may be associated with some plaster-based materials. SonaSpray is available in White, Arctic White, Black and custom colours. SonaSpray may be painted with a water-based paint in accordance with the manufacturer's instructions without significantly reducing the absorptivity.

K-13 samples with Protek protective surface coatings.

K-13 samples in a variety of custom colours.

COFFEE

HNK, Den Bosch, The Netherlands, by SKEPP design+build (2015). HNK provides flexible co-working space for independent consultants, free-lancers, telecommuters, small businesses, and travellers. The space is organised with a range of functions to serve different styles of work and the open plan concept is intended to promote networking and interaction. Dark grey K-13 has been spray-applied to the exposed metal decking and beams to provide uniform sound absorption throughout.

Hotel V Nesplein, Amsterdam, The Netherlands, designed and owned by Mirjam Espinosa (2013). The ground floor of the hotel is a large open space that encompasses the lively activities of a lobby, restaurant, and bar. The interior is designed to feel both cosy and luxurious, contemporary and vintage, boisterous and intimate. To help achieve the acoustic dimensions of these design objectives, custom-coloured SonaSpray cellulose was applied in a 12 mm thickness to the underside of the concrete slab ceiling in each space.

Keep Memory Alive Event Center Las Vegas, Nevada, designed by Gehry Partners (2010). The facility is part of the Lou Ruvo Center for Brain Health which offers integrated care, social services, and research for patients with brain disorders and diseases such as Alzheimer's and Parkinson's. While the clinic is dedicated to preserving memory, the Event Center was built for creating memories and is available for rent to the public with proceeds benefiting the Center. The space accommodates up to 400 seated or 700 standing patrons. The geometrically-complex interior is finished with Baswaphon acoustical plaster, controlling reverberation and providing clarity for the amplified sound system.

PLASTERS

Many acoustical plasters are in fact a thin layer of acoustically transparent finish on top of an absorptive backing panel such as fibreglass, mineral wool, foam, or recycled glass granulate like Reapor. The plaster finish is porous and allows sound to be absorbed by the backing material, and the absorptive performance may depend upon the thickness of backing material.

In the installation process, the backing panels are affixed to the wall or ceiling substrate and joints between the panels are taped or filled. A base coat may be trowelled or sprayed on and sanded, followed by one or more finish coats of plaster. The thin finish layer of plaster is typically composed of some granular aggregate such as glass or marble, which is highly porous once it dries and may require sanding between coats.

Most acoustical plaster systems can conform to curved, vaulted, and other complex geometries. The installation can be seamless for a surface area up to approximately 200-400 m^2. Larger areas will typically require expansion joints or transition surfaces to avoid cracking.

Acoustical plasters are aesthetically desirable because they appear to be a regular plaster or gypsum monolithic surface. Many plasters can be tinted or painted with manufacturer-approved paints. Installation often requires licensed installers and the multi-step process can be time-consuming and expensive.

BASWAPHON

The Baswaphon plaster system is comprised of a high-density mineral wool panel with a finished plaster coating composed of base layer of crushed marble aggregate which is spray-applied, gauged, and smoothed with a trowel. The surface finishes are available in Classic (smoothest, fine-grained finish), Resilient (washable, mould and mildew resistant, towelled smooth) and Frosted (washable, mould and mildew resistant, maintenance friendly, lightly sanded finish). The system is comprised of up to 95 per cent recycled content.

SONEX AFS

SONEX AFS by Pinta Acoustics is comprised of a melamine foam core panel which is laminated to a thin fibreglass mesh. The panels are applied to a solid substrate, the joints between the panels are taped and finished, and then two coats of Phonstop PA85 plaster are trowel-applied to create a smooth, porous, finished surface. The porosity of the plaster allows sound to be absorbed by the foam supporting structure.

SONAKRETE

SonaKrete is a cellulose-based plaster-like finish, similar in composition to K-13 and SonaSpray, but with a smoother and denser finish. Unlike many plaster products, SonaKrete does not use a porous backing material. During installation, finely ground cellulose is spray-applied in thin layers directly to the substrate and then hand-trowelled after each layer is applied. The finish is abuse- and impact-resistant, does not crack or spall with age, and expansion joints are not necessary. SonaKrete does not require extensive substrate preparation and is used primarily where drywall, plaster, brown coat, or any other smooth surface is present. It is available in White, Arctic White and custom-matched colours.

FELLERT

Fellert is comprised of a high-density fibreglass substrate with a spray-applied and hand-trowelled two-coat sound absorbing plaster, composed of cotton fibre and perlite. The finish texture is available in Secern (rough), Sahara (smooth plastered), and Silk (smooth).

Samples of StarSilent with Superfine (left) and Smooth (right) finishes on 25 mm thick granulated glass support boards.

STARSILENT

StarSilent System in the US, or StoSilent Distance in Europe, is a suspended panelised system with a smooth plaster finish. The backing panel comes in standard thicknesses of 10 mm, 15 mm, and 25 mm made of granulated glass from post-consumer recycled bottles. The system can form curved surfaces as well as compound curves. During installation, the backing panels are attached to a rigid metal framing with zinc coated drywall screws. The edges of the panels are glued together using a special adhesive, which is also spread over the screw heads. After light sanding, the panels are first coated with a base coat plaster. Once dry, the finish coat plaster is then hand applied to achieve a smooth and seamless finish. The plaster can be tinted or painted according to the manufacturer's specifications. The 'Superfine' finish option is a spray-applied plaster that visually appears smooth from distances greater than 5 metres. Acoustic performance is somewhat dependent upon providing an air gap behind the backing panels, and NRC ratings can range between 0.55-0.90. The StoSilent brand available in European markets offers a broader range of acoustical materials than is presently available in the United States and other regions, including direct-applied plasters and modular panelised products.

Pico Branch Public Library, Santa Monica, California, designed by Koning Eizenberg (2014). The building design explores the changing function of the civic library, from a place of quiet study to a community hub. The open plan layout and panoramic glass facade gives the feeling of being connected to the surrounding public park. The faceted ceiling is constructed of the StarSilent System, maintaining a sense of quiet in spite of the bustling activities.

TWA Flight Center at JFK International Airport, New York, New York, designed by Eero Saarinen (1962). An elegant and iconic landmark of neo-futurist mid-century architecture from the golden age of travel. The building's interior and exterior were meticulously renovated and brought up to code standards by Beyer Blinder Belle over several years and completed in 2011. As part of the renovation, the ceiling was refinished with Pyrok Acoustement 40 with a texture to match the original plaster.

PYROK ACOUSTEMENT

Pyrok Acoustement is a line of acoustical spray-applied plasters, similar in their application to stucco and traditional three-coat plasters.

For interior applications, Acoustement Plaster 20/40 are gypsum-based plasters with exfoliated vermiculite that are typically applied to drywall substrates and may be trowelled to a semi-smooth finish or painted. Acoustement Plaster 40 has a denser formulation than 20 and is intended in areas requiring abuse resistance.

Acoustement 40 is a Portland cement-based plaster with exfoliated vermiculite that is highly abuse-resistant and can be used in interior or exterior applications and harsh environments such as traffic tunnels, correctional facilities, gymnasia, and manufacturing facilities. This plaster may double as a fireproofing material and may be coloured with iron oxide pigments.

Samples of Pyrok Acoustement Plaster 20 (left) and Acoustement 40 (right).

Deloitte Touche Tohmatsu, Copenhagen, Denmark designed by 3XN (2005). Artificial and natural lighting are used in a variety of innovative ways throughout the office building. In the company's screening room, lighting above the translucent TechStyle panels provides a uniformly diffuse ceiling surface. TechStyle's swing-down clip suspension system allows easy access to the plenum for replacing light bulbs.

TECHSTYLE

Techstyle by Hunter Douglas Architectural is a unique panelised system composed of a formaldehyde-free non-woven fibreglass textile with a honeycomb core construction. The panels are incredibly lightweight, less than 1.5 kg/m², and capable of spanning as much as 1.5 square metres without sagging. Their light weight makes installation easier and shipping costs lower.

Techstyle panels are designed for walls or ceiling installations. Standard lay-in grid systems can be used or, for a more monolithic appearance, the panels have a specially designed semi-concealed system that provides hinged, swing-down access to the ceiling plenum.

A variety of aesthetic finishes are offered with the Techstyle system: solid colours, graphic patterns, and textured panels to create the appearance of stone, leather, stucco, plaster, or wood finishes. Some of the panels are semi-translucent, capable of providing diffuse light when backlit.

The panels are rated NRC 0.85 when mounted in a suspended ceiling (E-400) and achieve CAC-17 without the use of any additional backing material.

Techstyle samples in Wood, Graphic, and White.

Primacoustic and Terzacoustic transparent fabrics with a matte finish by Creation Baumann.

TRANSLUCENT ACOUSTICAL TEXTILES

For most textiles, a high degree of acoustic absorption is a function of density and thickness, such as a heavy velour curtain. Sheer or translucent textiles have historically been acoustically transparent, absorbing very little sound. In recent years, translucent acoustical textiles have been developed using Trevira CS, which is a durable, inherently flame resistant, polyester fibre. Clear polyester fibres provide light translucency, while the tightness of the weave can create a high flow resistance that provides sound absorption. Some products, such as WeavePerf by Akustik & Innovation, utilize micro-perforations to further enhance sound absorption.

Translucent acoustical textiles can be installed in a number of ways but are most commonly used as drapery, allowing for variable acoustical control in environments such as conference rooms with glass walls. As with any textile, the absorptive performance will also depend upon the fullness of the hung drapery and the depth of spacing away from the wall. The producers of these textiles include Creation Baumann, Annette Douglas' Silent Space Collection distributed by Vescom, and Akustik & Innovation.

Samples of AlphaAcoustic transparent fabrics by Creation Baumann, available in 16 standard colours.

Private residence in St. Gallen, Switzherland, featuring Zetacoustic fabric from Creation Baumann.

Batyline Aw samples in standard colours, left to right: Bamboo, Alaska, and Lux.

BATYLINE AW

Batyline Aw, manufactured by the French group Serge Ferrari, is a micro-textured extruded polyester mesh acoustical textile that can be installed in a variety of manners — as tensioned sails, stretched ceilings, banners, curtains, shades, or as diffusely rear-illuminated surfaces. It is resistant to deformation, tearing, impacts, and abrasion and can be installed in complex shapes, curves and very large unsupported spans. The textiles are translucent and may be installed as light filtering shades or coverings for skylights. When installed with an air gap of 100 mm or more, Batyline Aw is shown to achieve an NRC rating of 0.65; higher NRC ratings can be achieved by using a porous backing material. Batyline Aw is fully recyclable, GreenGuard Gold certified, and can be digitally printed upon, and is available in 12 standard colours.

SILENTPROTECT

SilentProtect is designed to be installed in tension like a sail with an air gap between the fabric and boundary surface. The textile is made from a polyethersulfone (PES) thermoplastic, woven like a textile, and then coated with PVC, giving it dimensional stability, tensile strength, and durability. It can be fabricated in square or triangular shapes with large spans that make it well suited for large volume spaces such as lobbies, swimming pools, and sports arenas. It is moisture resistant, dirt-repellant, and semi-translucent.

Klosterkirche, Dargun, Germany, designed by Beyer Architekten (2014). The historic Cistercian monastery chapel was modernised to allow for a variety of events and performances in the nave of the church. Batyline Aw was used to create 10 metre sails that are attached to an aluminium framework, formed to reference the original geometry of the Gothic vaulted ceiling. Baytline Aw was a lightweight solution for providing uniform sound absorption and a light reflecting surface for different lighting conditions.

Soundtex®

SOUNDTEX

SoundTex is a non-woven fleece made of cellulose and glass that is 0.2 mm thick and commonly used as a backing material for perforated metal, wood, and gypsum. SoundTex is specially designed with an optimised flow resistance so that sound is absorbed when passing through the material. The material is still air permeable, which may be beneficial for reducing moisture in ceiling plena, providing flexibility in HVAC design, and allowing for thermal transfer in energy-efficient chilled ceilings.

The material is available in custom and standard colours — black, light grey, dark grey, and white. It is easy to cut and can be bonded or adhered to a variety of surfaces. In fact, the manufacturer can provide SoundTex with a pre-applied heat-activated adhesive.

When used as a backer for perforated panels, the acoustic performance will depend upon the perforation open area, the thickness of the panel, and the mounting depth or air gap distance. The use of SoundTex may eliminate the need for a bulky porous backing material such as fibreglass or mineral wool. In such configurations, the manufacturer recommends an air gap of at least 200 mm behind the perforated panel, which can typically yield an NRC rating of around 0.75.

SoundTex is sold in master rolls (nominal 1800 mm wide x 450 linear metres), custom rolls, or in custom-cut sheets. SoundTex should not come in direct contact with moisture and installations in humid environments, such as swimming pools, should be avoided.

Opposite Above: Standard black SoundTex application, showing the speckled texture of a pre-applied heat-activated adhesive for bonding.

Opposite Below: An assortment of standard and custom colours.

The University of Wisconsin's McClain Athletic Facility, designed by VOA Associates (2014). The field's 3800 m² roof is created with an insulating and translucent Tensotherm system with a Lumira aerogel core.

TENSOTHERM

Tensotherm is a tensioned membrane composite system that is used for lightweight, long-span roofing systems such as athletic facilities and stadiums. The material is comprised of a translucent blanket of aerogel insulation that is sandwiched between two layers of polytetrafuoroethylene (PTFE)-coated fibreglass. Aerogel is at the heart of Tensotherm's many attributes: thermal insulation, transmission and diffusion of natural daylight, and acoustic absorption.

FBC Office Tower lobby and cafe renovation, Frankfurt, Germany designed by Just /Burgeff (2007)
The Lightframe system creates a luminous canopy that spans between the foyer, porte cocher, and cafe. Skylights allow daylight to illuminate the translucent textile membranes with varying intensity throughout the day according to the position of the sun. At night, the ceiling softly glows with artificial back lighting.

SEFAR Architecture fabric IA-80-CL (uncoated), left; SEFAR Architecture fabric IA-85-OP, (coated and micro-perforated), right.

SEFAR LIGHTFRAME

Sefar produces a number of architectural fabrics which are suited to both interior and exterior applications. Lightframe is a modular system that has been designed to provide both translucent light diffusion and sound absorption. The Lightframe system is comprised of an upper and lower textile membrane held in tension by a lightweight extruded aluminium frame with gasket seals along the perimeter. The textiles allow for more than 75 per cent of light transmission and the even diffusivity of the material gives the impression that the fabric itself is glowing.

Lightframe is used with one of two available polyvinylidene fluoride (PVDF) textiles: Sefar Architecture IA-80-CL is an uncoated fabric for interior applications with a weave structure specially designed to absorb sound; and Sefar Architecture IA-85-OP is a micro-perforated PVDF fabric coated with a fluropolymer mixture that allows it to be used in exterior applications.

All fabrics in the SEFAR Architecture range are ASTME E 84 Class A rated for flame-resistance, producing very little smoke and no burning droplets. The fabric is VOC-free and can be printed upon. The acoustic performance of the Lightframe system will depend upon the particular textiles and the mounting conditions. Test reports have indicated the system is capable of achieving NRC ratings up to 0.90.

FACINGS

OVERVIEW

Facings are often used protect porous absorbers from damage, moisture, or contamination; they may prevent friable fibres from coming loose or coming in contact with skin; they may allow for an absorptive surface to be regularly cleaned; or they may simply hide the porous material from view and create a different aesthetic appearance.

A solid material can be fabricated to be acoustically transparent, or transondent, simply by adding perforations, cutting slits, or arranging the material with gaps provide sufficient open area to allow sound unimpeded access to a porous backing material. For very thin rigid materials, such as sheet metal, it is generally recommended to have an open area of 20 per cent or more. For thicker materials, such as wood slats, the open area requirement may be much higher — 30, 40, or 50 per cent. Transondent coverings can be made from a variety of materials such as perforated sheet metal, expanded metal, welded wire, timbre slats, or open masonry units. As the thickness of the facing material increases or as the percentage of open area decreases, then the facing material will become less transparent and will reduce access to the porous absorber, particularly at high frequencies. When this occurs the facing and absorber may begin to behave more like Helmholtz absorbers, as discussed in the following chapter.

Pliable or flexible materials, such as fabrics or films, are also commonly used to cover porous materials. As a general rule of thumb, if you can hold a textile to your lips and easily blow air through it, it is acceptable to use as a covering and will not have a significantly negative impact on the sound absorption of the underlying porous material. Impervious textiles or films, like leather, vinyl, plastic, or coated fabrics may require perforation to allow for airflow. There is no standardised method for quantifying the acoustical transparency, or transondence, for these applications, although some experimental methods have been proposed.

There are many work environments that require surfaces to be cleanable and non-porous so they do not retain dirt or dust particles. These spaces include clean rooms, food preparation areas, micro-electronics facilities, and operating rooms. Acoustic absorption is desirable to maintain appropriate ambient noise levels and improve speech intelligibility for the occupants. In these environments, porous materials are sealed or wrapped with thin impervious materials which do not allow air or moisture through. This impervious layer will limit the ability for sound to access the porous material, reducing the sound absorption performance. To limit this reduction of performance, it is recommended that this impervious layer be very thin (< 2 mil) and relatively limp — not rigidly adhered to the porous material. As shown in the figure, sound absorption losses occurs mostly at high frequencies and the losses increases with the increasing thickness of the encapsulating material.

A few ceiling tiles are designed for such specialised environments, which are fully encapsulated with a thin film that protects the porous absorber and allows the surface to be fully cleaned. The International Standard EN ISO 14644-1 standard (classes 1 to 9) is used for the classification of air cleanliness. This is the official standard which has replaced the US Federal Standard 209E (classes 1 to 100,000). For example, a Class 5 Clean Room means that fewer than 3520 particles (0.5 microns in size) are present per cubic metre, which equals 100 particles per cubic foot. The pharmaceutical industry is regulated by Good Manufacturing Practice (GMP) standards (Class A to D). Ceiling tile manufacturers will typically indicate which class their product falls under. In many cases, tiles will be installed with hold-down clips to seal the tile into the T-bar grid. This also allows the tiles to be washed without popping the tile out of the grid.

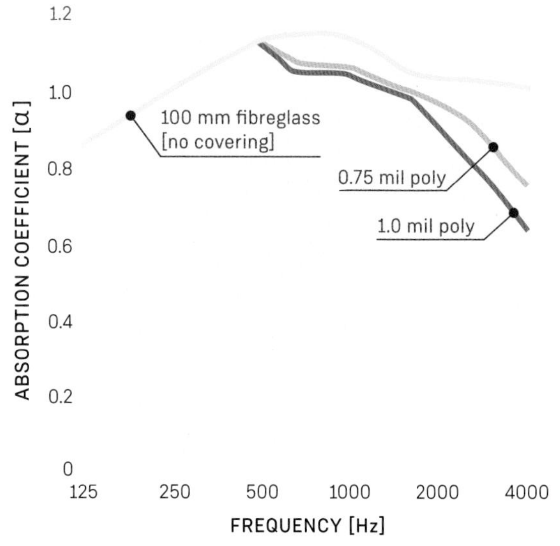

Data from three test reports of 100 mm fibreglass with two different thicknesses of polyethylene film coverings. As the film thickness increases, the mid and high frequency absorption performance declines.

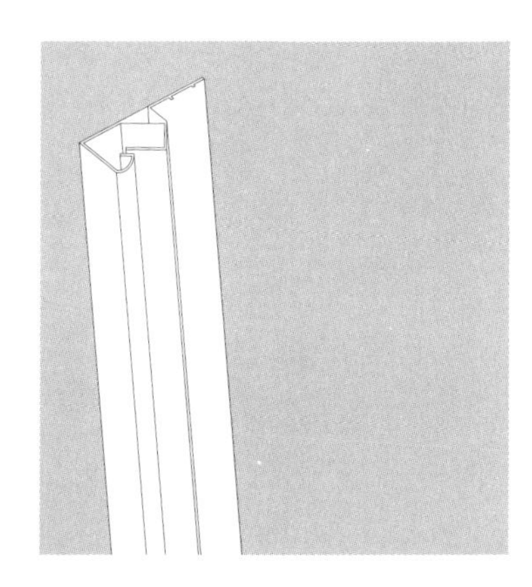

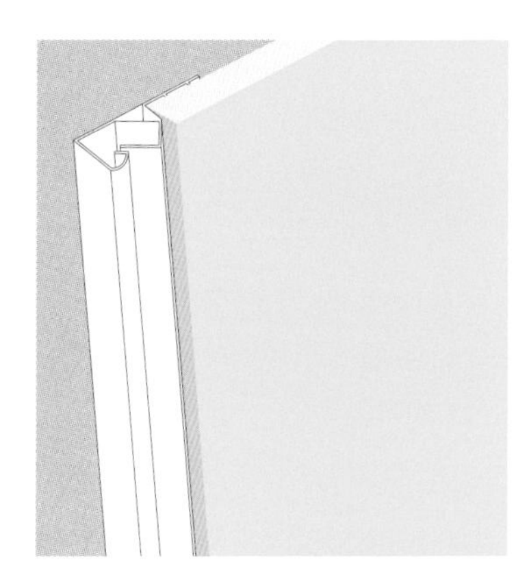

Stretched fabric systems consist of a track or framework attached to the wall or ceiling and a porous absorber installed between the tracks. Fabric is stretched over the porous absorber and tucked into the interlocking jaws of the tracks.

STRETCHED FABRIC SYSTEMS

Stretched fabric systems consist of three elements: a framework, an acoustical core or substrate material, and fabric. The frames or tracks are directly mounted to the wall or ceiling. An acoustical material is directly applied between the tracks - the depth of the tracks is selected to match the acoustical material thickness. Finally, fabric is stretched over the framework and tucked into the locking harpoon jaws of the frames, typically without the use of adhesives or mechanical fasteners. Fabric can usually be easily removed and replaced while leaving the framework and acoustical backing material in place.

The frame or track hardware may be plastic or metal, and a variety of profiles such as Square, Bevel, or Radiused can provide different edge finish effects. Most manufacturers provide additional framework details to accommodate light fixtures, electrical outlets, fire sprinklers, or HVAC fixtures so that the finished installation is fully integrated and seamless.

A variety of fabrics can be used. Typically polyester and polyolefin fabrics work well; cotton, silk, and wool fabrics may require additional stabilizing; nylon, rayon and viscose fabrics can be problematic to stretch. The fabric needs to be acoustically transparent in order for it to allow sound to access the absorptive backing. As a rule of thumb, if you can hold the fabric to your lips and blow air through it with relative ease, the fabric is likely acceptable. Extra-wide fabrics are often used to avoid seams and create the appearance of a monolithic surface. The advent of digital printing has also opened up a number of aesthetic possibilities for stretched fabric such as art, murals, advertising, or trompe l'oeil architectural effects, serving a double function as acoustical absorption.

The type of backing material and its thickness will dictate the acoustic absorption performance. Common materials include fibreglass, cotton, or polyester. High density backing materials can sometimes double as bulletin boards.

This large corporate conference room, designed by HLW, uses stretched fabric on the walls and ceiling, provided by Fabric Systems. The fabric creates the appearance of a monolithic surface and the tracks integrate seamlessly at various angles with overhead lighting. The acoustical backing prevents flutter echo from the panoramic windows and assists speech intelligibility for meetings and teleconferencing.

Architextile, Acoustic Textile Panels, designed by Aleksandra Gaca. With a passion for three-dimensional structures, textile designer Aleksandra Gaca has developed a series of woven textiles for acoustic panels which can be used as wall coverings, suspended panels, or free standing room dividers. The fabrics are created at the intersection of technology and handicraft, using a patented weaving technology that produces the three-dimensional surface effect. The fabric is a mixture of wool, mohair, and metal threads. Custom colours and compositions are possible. The raised textured surface may provide additional absorption at very high frequencies but is particularly elegant in its effects of light and shadow.

Black, white and red PVC vinyl with heat-sealed seams around a fiberglass core manufactured by MBI.

PLASTIC FILMS

Thin plastic films are frequently used to encapsulate porous materials for use in harsh or humid environments. Plastic films can be manufactured very thin — beneficial for maintaining acoustic performance. Many of these films are thermoplastics, which means they can be reprocessed and bonded together using heat. The most common films currently used for acoustical material coverings are PVC, PVF, and PE.

Polyvinyl Chloride (PVC), known commonly as vinyl, is water resistant and has inherent fire retarding properties due to its chlorine content. It is also resistant to acid, alkali, and most inorganic chemicals. PVC films can be made in clear, opaque, and a variety of colours.

Dupont Tedlar Polyvinyl Fluoride (PVF) is structurally similar to PVC and ideal for harsh environments because it is UV resistant, non-stick, easy to clean, and resistant to weathering and staining.

Polyethylene (PE) is the most widely used plastic in the world, with products ranging from clear food wrap and shopping bags to detergent bottles and automobile fuel tanks. It can also be slit or spun into synthetic fibres or modified to take on the elastic properties of a rubber.

Acoustical Flexible Duct is an excellent example of a ubiquitous, encapsulated sound absorbing material. An outer layer of fibreglass provides thermal insulation for the conditioned air, while a very thin layer of nylon, mylar, or polyethylene, keeps fibres out of the airflow and allows for sound attenuation through the ductwork.

NYLON RIPSTOP SAILCLOTH

Nylon is a generic term for a family of synthetic polymers. It is a silky thermoplastic that can be melt-processed into fibres, films, or forms. It is waterproof and resistant to mould, insects, and fungi. Although reasonably resistant to many everyday substances, nylon will dissolve in phenol, acids, and some harsh chemicals.

Ripstop Sailcloth has large diameter threads for reinforcement woven into nylon fabric at regular intervals creating a pattern of small squares. Small holes or rips will stop once they encounter the first large diameter thread. The material was developed during World War II as a replacement for silk in the production of parachutes. You can find nylon ripstop in a variety of colours and weights or thicknesses. Textures range from soft and silky to crisp and stiff.

Blue ripstop nylon with sewn edges wrapped around a fibreglass core board manufactured by MBI.

HOME
GUEST

Raritan Valley Community College Natatorium, Somerville, New Jersey, designed by Peter Johnston (2005). Prior to renovation, the natatorium was a concrete box that suffered from poor acoustics and inadequate lighting. Encapsulated lapendary panels were hung in a catenary fashion across the ceiling, improving the acoustics and providing a visual alteration of the geometry of the space. The suspended linear light fixtures use a single bulb at one end, which allows bulbs to be changed without the need for scaffolding.

The Seattle Public Library, Seattle, Washington, designed by OMA (2004). The library has been designed with a variety of functions and corresponding acoustic environments. On the tenth floor, the library's reading room offers views of the cityscape and Elliot Bay. The atmosphere is bright, quiet, comfortable, and soft due to a pillowy ceiling composed of acoustical absorbers covered in ripstop nylon. The pillows have been designed as a grid, providing an organizing structure for lighting, sprinklers, smoke detectors and other building services.

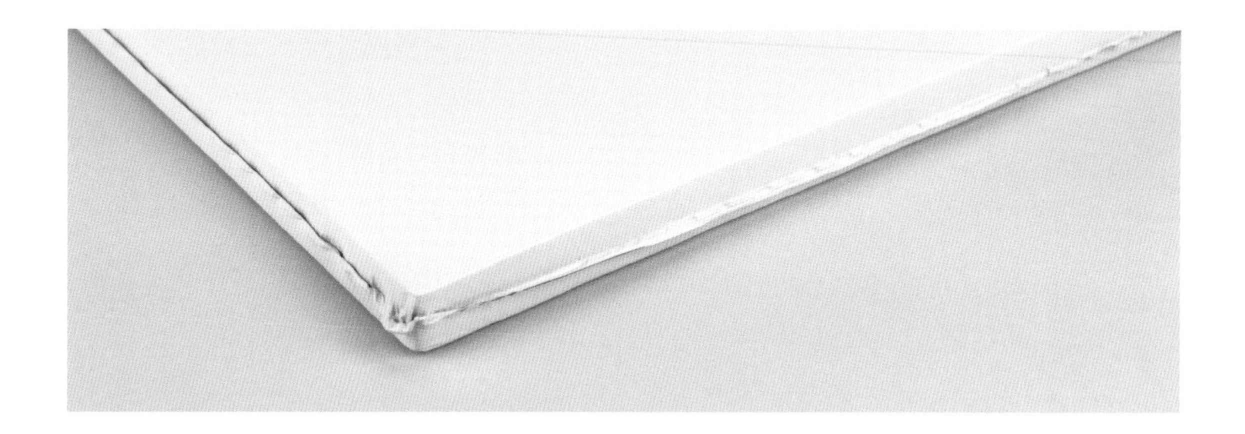

ECOPHON HYGIENE ADVANCE

Ecophon Hygiene Advance is composed of a high-density, third generation glass wool core that is fully encapsulated on all sides with a smooth high-performance Teflon film. The tile is dirt-repellent, impervious to particles and water, and resistant to most chemicals. 20 mm thick tiles mounted with a 200 mm air gap (E-200) achieves NRC 0.80.

Ecophon Hygiene Advance A Ceiling Tiles are used to improve speech intelligibility in an operating room.

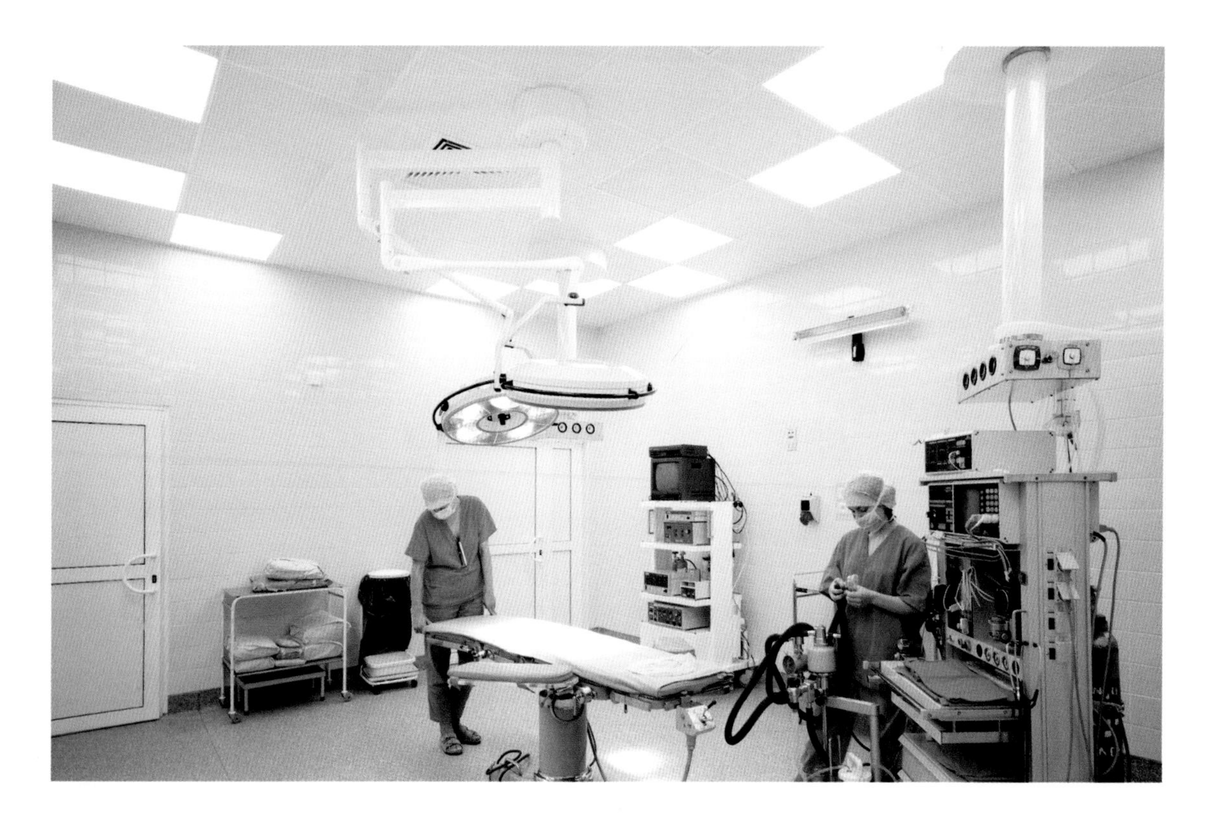

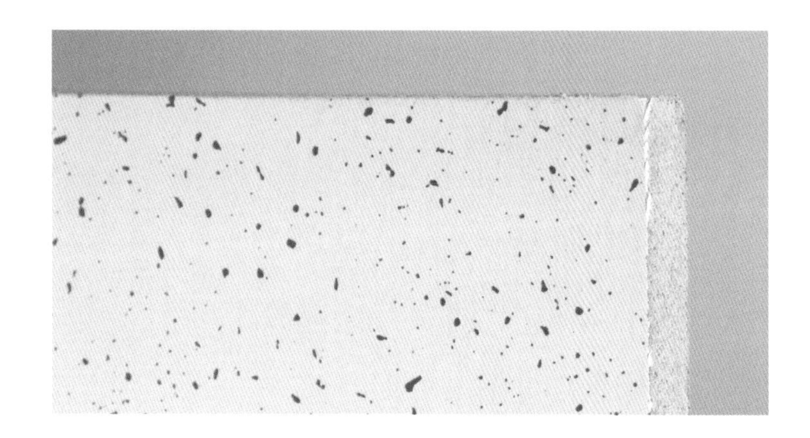

ARMSTRONG CLEAN ROOM FL

A wet-formed mineral fibre tile suitable for food preparation areas and up to Class 100 clean rooms. A thin soil-resistant, washable, polyester film covers the face of the tile. The film is adhered to the edges of the tiles, leaving most of the surface area unadhered. Absorption is provided through diaphragmatic action of the film facing, achieving NRC 0.55 with an E-400 mounting.

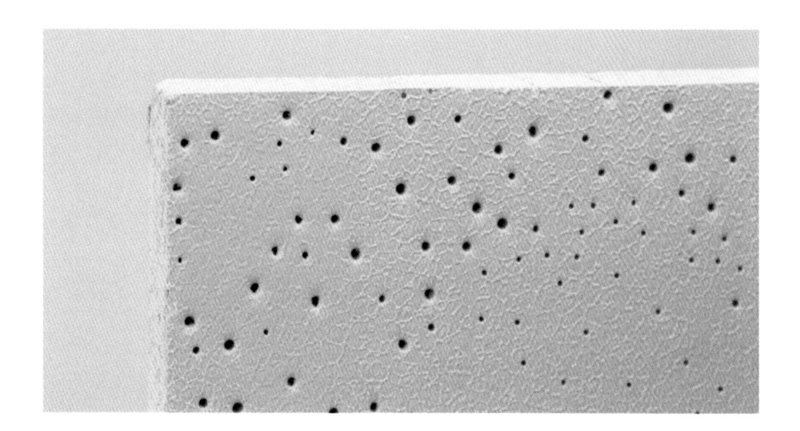

USG CLEANROOM CLIMAPLUS

The CleanRoom ClimaPlus 10M-100M ceiling tile has an embossed, vinyl-laminated face with sealed back and edges, suitable for use in Class 10M-100M (ISO 7) clean rooms. Perforations puncture the vinyl facing, allowing sound to absorb into the wet-formed mineral fibre tile, yielding an NRC rating of 0.55 in a suspended ceiling [E-400]. The tiles contain more than 50 per cent total recycled content and may be easily cleaned a damp sponge, mild detergent, and water.

VACU-BOND

Vacu-Bond is a process developed by Acoustic Enterprises that allows the re-use of old or soiled fibreglass ceiling tiles. The existing tiles are inserted into a vacuum-sealed bag which has a washable white nubby fabric finish on the front side, vinyl reinforced foil finish on the backside. The washable feature makes this product a viable option for clean rooms or healthcare spaces. The product advertises a 40-50 per cent cost savings over replacement and an added 15 or more years to the lifecycle of the existing fibreglass tiles. Tests have shown the tiles can achieve NRC 0.85 with E-400 mounting.

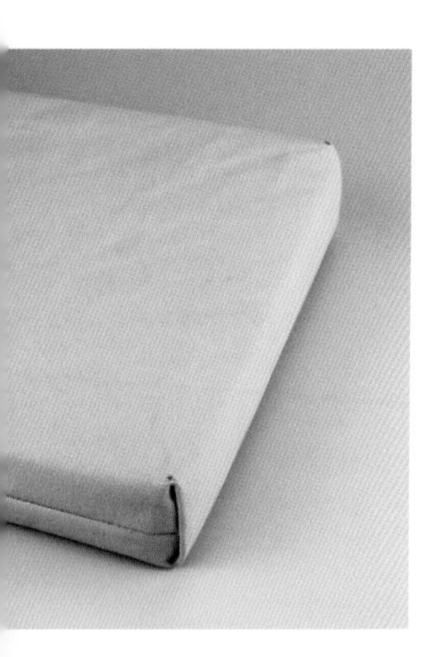

MBI WEATHER RESISTANT EXTERIOR PANELS

Low-density fibreglass is wrapped in a proprietary 'Cypress' exterior fabric that allows the panels to withstand the elements and temperature changes outdoors. These panels are typically mounted with stainless steel grommets or an aluminium stiffener/gripper bar. The fabric is available in 19 standard colours, pigmented and then top coated with acrylic. The panels are available solid or perforated. When the outer fabric is perforated, a PVC liner is used to protect the inner fibreglass core from moisture.

VAPOUR RETARDERS

Vapour Retarders are used to limit moisture from penetrating into thermal insulation; commonly used to hold batt insulation between joists, laminated directly to fibreglass or mineral wool boards or batts, or used as protective wrappings for encapsulated acoustical panels. Pictured here are samples manufactured by Lamtec, each composed of a thin film laminated to a foil or kraft paper backing with fibre reinforcement.

Lamtec vapour retarding materials, from left to right : (30J White All Service Jacket) elastomeric polymer film with aluminium foil backing; (WMP-VR-R Plus) polypropylene scrim with metalised polyester backing; (WMP-VR) black polypropylene film kraft scrim backing, (R-3035 HD) aluminium foil with elastomeric polymer coating and kraft scrim backing.

QUILTED ABSORBERS

Quilted absorbers are fabricated from of batting of various thicknesses, sewn between two layers of facing material. They are low-cost, flexible, easy to install, and depending upon the facing material, may be cleanable and resistant to moisture, humidity, dirt, debris, oil, grease, mild acids, alkalis, and high temperatures.

Common facings are plain fibreglass cloth, grey aluminium-vinyl coated fibreglass cloth, silicon coated fibreglass cloth, or 100 per cent spun-bonded polyester scrim. When the facing materials are non-porous, such as coated fibreglass cloth, sound absorption occurs diaphragmatically where each quilted square functions like a diaphragm that transmits sound energy from the non-porous facing to the core batting material. Thicker quilted absorbers will increase low-frequency performance but non-porous facing may reduce high-frequency absorption.

Quilted absorbers are used for a variety of industrial and OEM applications including mechanical rooms, equipment enclosures, machinery housing, back-of-house theatre spaces, swimming pools and gymnasia, and aerospace linings. The products are provided as rolls, die cut shapes, or custom-sized panels. The material is easy to work with and often mounted using grommets, industrial Velcro, adhesive, or a variety of mechanical fasteners.

Blankets are commonly adhered to mass-loaded vinyl to provide both sound absorption and sound transmission loss. This is particularly useful when erecting a barrier around a noise source, such as a temporary barrier around construction equipment.

Sound stages often use quilted absorbers on walls and ceilings to create an acoustically dead space so that sound, such as dialogue, can be clearly recorded along with the image for film and television productions. The fibreglass quilted absorbers pictured here at Santa Fe Studios, Santa Fe, New Mexico were manufactured by Insul-Quilt.

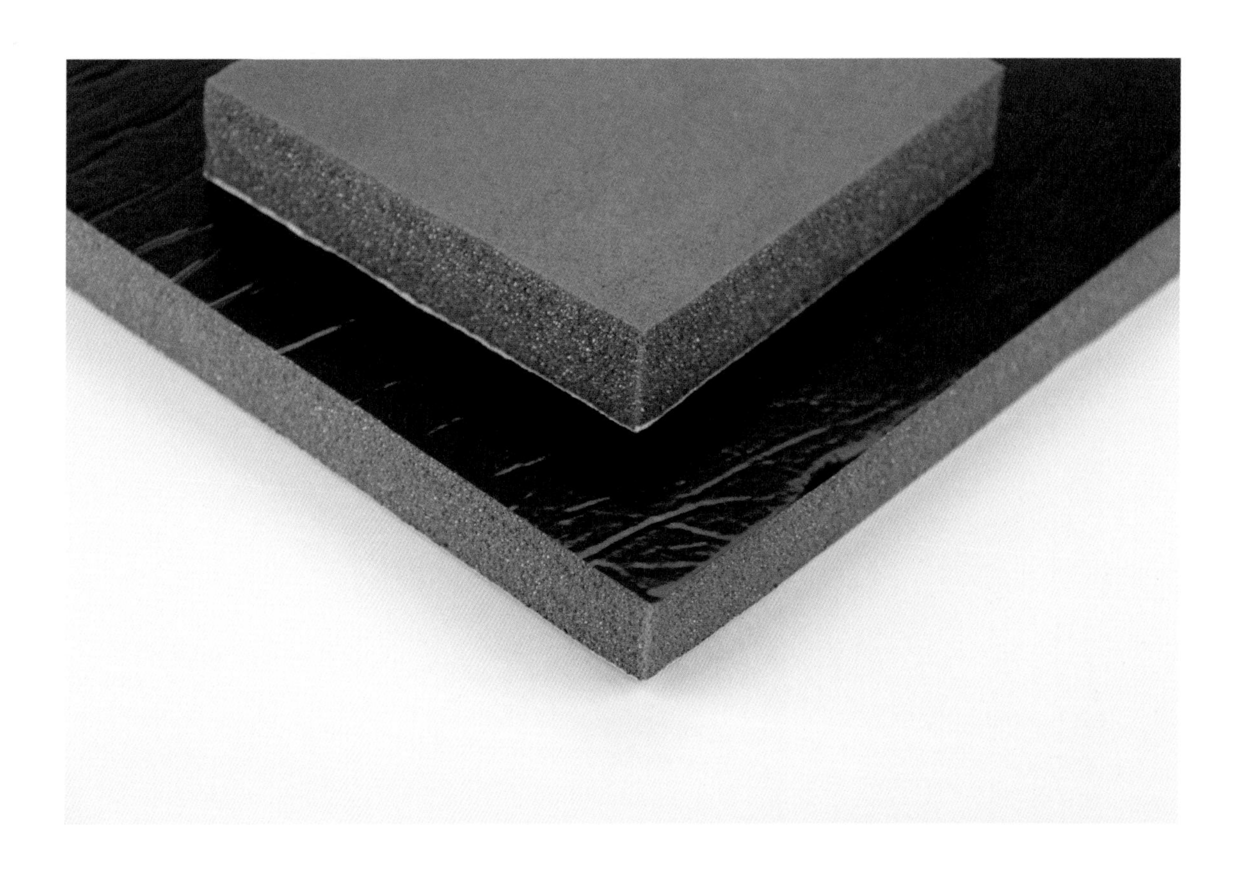

FOAM FACINGS

Flexible open-cell foams, such as polyurethane or melamine, are shapeable, bendable, lightweight, and relatively inexpensive. These foams find a variety of noise-reduction applications as linings inside the motor housings of machines and everyday appliances, inside the operator cabins of heavy machinery, or inside vehicles. In order to protect the foams from abrasion, dirt, moisture, heat, and chemicals under these harsh conditions, foams are often bonded or laminated to thin facing materials.

Porous facings include fabrics or perforated films, such as perforated vinyl, which allow sound to penetrate the foam and may not have a significant reduction on the absorption performance. Foams can also be densified, using compression under high heat to create a thin skin of denser surface that provides some additional stiffness and abrasion resistance without fully sealing the porosity of the foam surface.

Non-porous facings will alter the absorptivity of the foam, typically reducing high frequency absorption and enhancing some mid or low frequencies. These facings are thin films which may have fibre reinforcements. Common films include polyester, urethane, PVC vinyl, Mylar, Tedlar PVF, or aluminium foil.

Opposite Above: TUFCOTE foams with densified textured surface,top, and 2 mil thick black urethane facing, bottom.

Opposite Below: TUFCOTE foams with 1 mil solid, perforated, and reinforced aluminium foil facings.

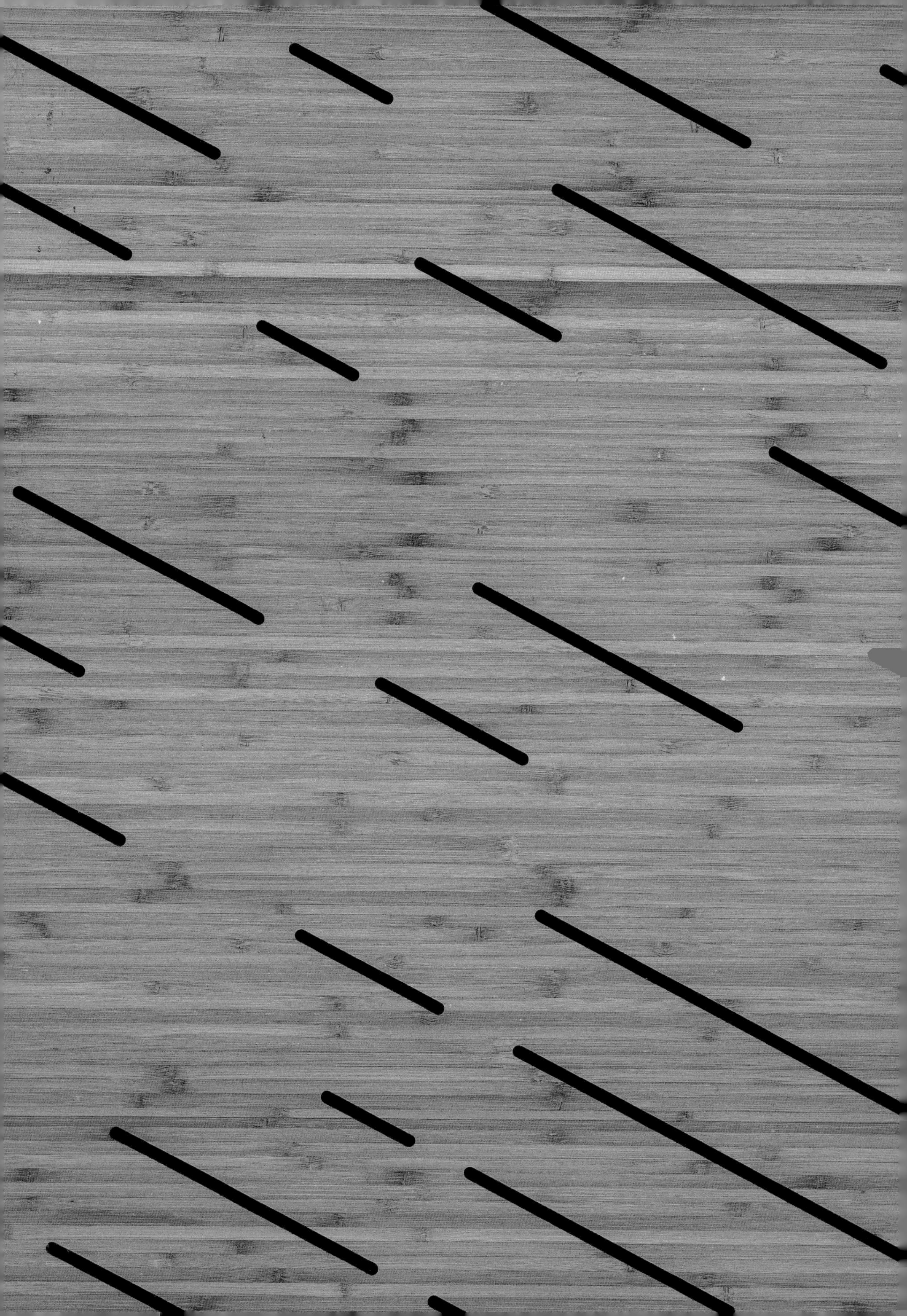

RESONANT

OVERVIEW

There are two types of sound absorbers which absorb sound through resonance — Helmholtz absorbers, which are named after the physicist Hermann von Helmholtz, and membrane or panel absorbers. These types of absorbers function differently from porous materials and tend to absorb over a smaller range of frequencies.

Helmholtz absorbers can take many forms and are often referred to by different names - cavity absorbers, volume absorbers, perforated resonators, or slit resonators. The easiest way to experience a Helmholtz resonator is to blow across the top of the bottle and listen to the tone produced at its natural resonant frequency. A Helmholtz resonator is simply an enclosed volume of air with a short open neck. When sound waves strike the opening in the neck of the resonator, they set the air within the neck in motion. The air inside the neck behaves like a solid plug that moves back and forth, compressing and expanding the air inside the enclosed volume like a spring. As the air moves back and forth, the acoustic energy is absorbed — converted to heat as a result of the friction of this movement. To improve the absorption of this system, damping is necessary and this is achieved by placing a porous material within the backing cavity and near the neck of the resonator.

With resonant absorbers, maximum sound absorption occurs at the peak resonant frequency of the entire system and the absorptivity falls off like a bell curve above and below this resonance. The following formula approximates the resonant frequency of a simplified Helmholtz resonator and shows the relationship between the dimensions of the neck and the volume of the cavity.

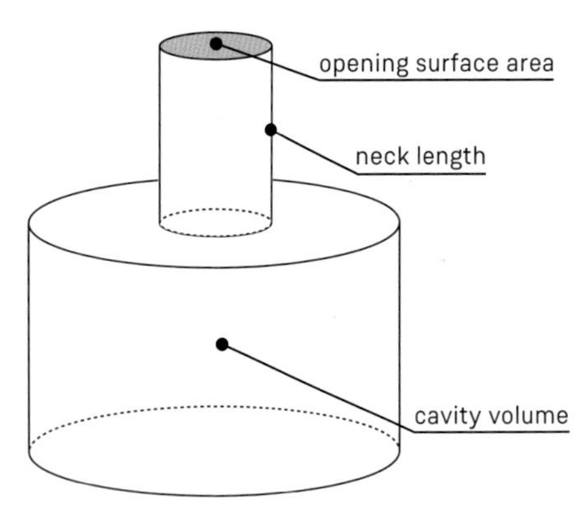

The basic form of a Helmholtz resonator.

$$f = \frac{c}{2\pi}\sqrt{\frac{S}{LV}}$$

f = resonance frequency (Hz)
c = speed of sound (343 m/s)
s = cross-sectional area of the neck (m^2)
l = length of the neck (m)
v = volume of the cavity (m^3)

The addition of porous material to the Helmholtz resonator can increase the peak absorption value and broaden the bandwidth or range of frequencies of absorption.

Acoustical masonry units are a common example of a Helmholtz resonator; each unit exhibits the basic form of an opening, neck, and cavity. Helmholtz resonators can also be constructed from perforated, milled, or punched openings in panels, or by creating an open area between spaced material like wood battens. In these cases, the perforations or slit openings function like a series of tiny bottle necks and the space behind the panel functions as the cavity.

By altering the length or diameter of the neck or changing the volume of the air cavity of Helmholtz resonators, it is possible to select the desired resonant frequency or range of frequencies. With these adjustable parameters, materials can be designed to tackle specific noise problems. A common application is for noise reduction of electrical transformers, which produce tonal noise at 120 Hz. The transformer's metal enclosure may be designed with interior perforated metal to absorb sound at that specific frequency. Alternately, the electrical room which houses the transformer may be constructed with acoustical masonry units that are specially designed to absorb sound at 120 Hz.

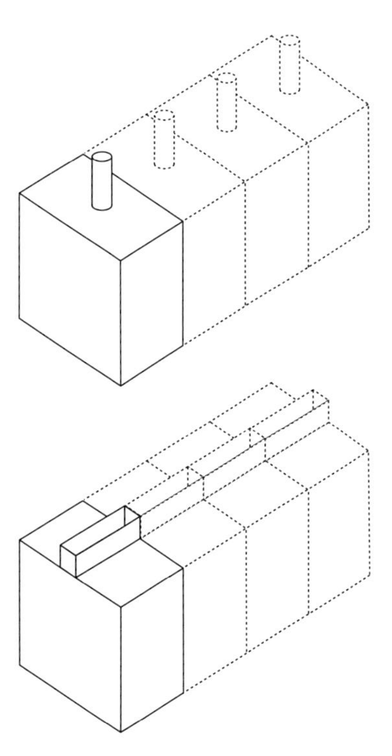

When a panel is perforated, it creates a series of small Helmholtz resonators where the perforation forms the opening, the thickness of the panel forms the neck, and the air space behind the panel constitutes the cavity volume.

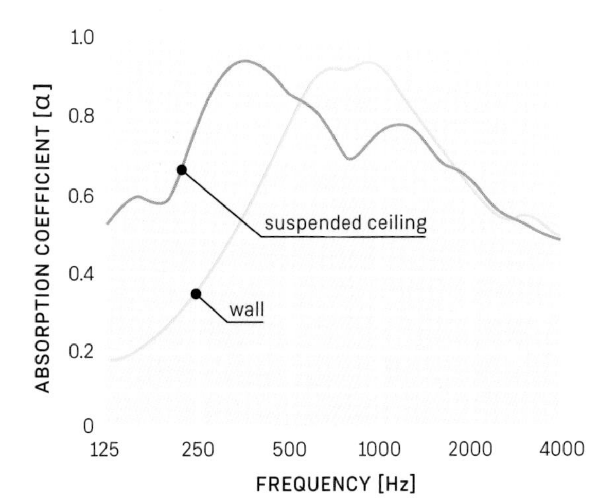

The measured absorption coefficients for the same perforated wood Helmholtz-style absorber when mounted with a 50 mm gap against the wall compared to a suspended ceiling [E-400] mounting. As shown, the peak resonance has shifted to a lower frequency due to the increase in cavity depth.

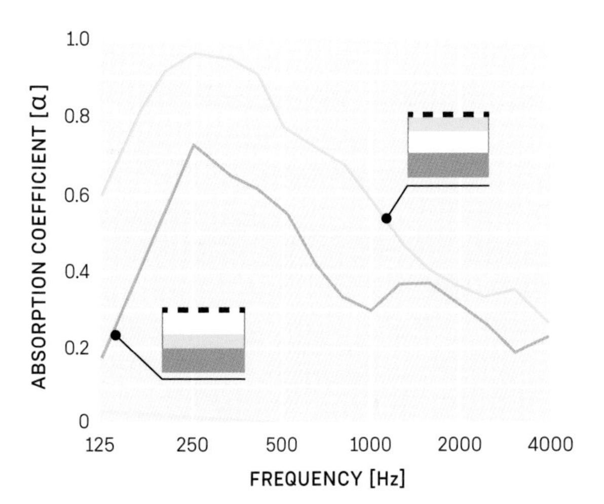

The measured absorption coefficients for the same perforated sheet metal panel with the porous absorber placed directly against the sheet metal compared to the condition where the absorber is mounted against the wall and spaced away from the sheet metal. Performance is dramatically improved when the porous material is installed directly against the sheet metal.

When using panelised Helmholtz absorbers, changing the mounting depth of the cavity will result in a change to the peak resonance frequency because the effective volume of the resonator is being altered. For systems mounted against a wall, the mounting depth may only be 25-50 mm. When that same material is mounted as a suspended ceiling, the mounting depth may be 400 mm or more and the resonant frequency can shift down considerably. When consulting a product's acoustical test data, it is recommended that the test conditions closely align to the install condition.

Perforated materials can acoustically transparent if enough percentage open area exists but will begin to function as Helmholtz style absorbers when the open area is reduced. To achieve the best absorptive performance, the porous backing material should be placed directly against the perforated panel. While it is sometimes easier and convenient to install porous materials against the wall, the performance is significantly enhanced whenever spaced away from the reflective surface, as explained in the porous materials overview.

Another factor that can impact absorption performance is the flow resistance of the porous backing material. As discussed in the porous materials overview, the flow resistance refers to how easily air can move through the material. For fibrous materials there is a relationship between fibre density and fibre diameter, such that the denser the material and the finer the fibres, the higher the flow resistance; as well, the thicker the material, the higher the flow resistance. When installing a Helmholtz absorber type product, it is good practice to use the same or equivalent type of porous material that the manufacturer used when testing their product so there are no unexpected variations in the absorption performance of the installed material. For example, if a manufacturer sells a perforated wood product that was tested using a 25 mm thick 16 kg/m³ density fibreglass backing, then the same thickness and density of fibreglass material should be used when installing the product. Using a different density or thickness of backing material may unintentionally alter the absorption performance.

Designing and optimising a new type of Helmholtz absorber is rather complex and will likely require the assistance of an acoustical engineer. However, the basic parameterisation of Helmholtz absorbers allow for a vast variety of materials and finishes to be transformed into sound absorbers which may be finely tuned to suit the specific needs of each application.

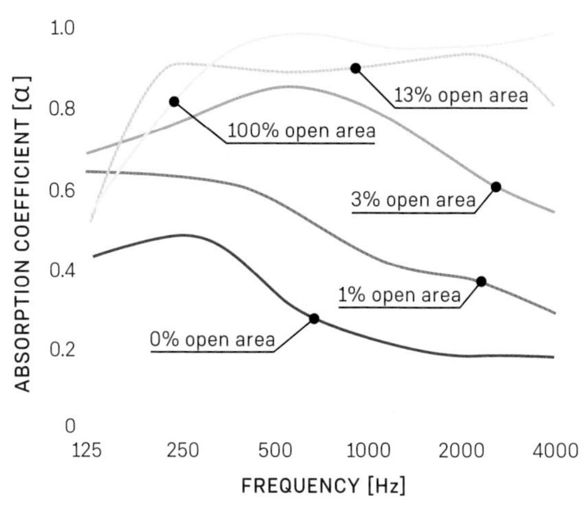

The sound absorption coefficients for perforated panels with porous backing materials. The figure demonstrates the ways in which the percent of perforation open area can alter the absorption performance. Thirteen per cent open area is somewhat transondent but begins to reduce high frequency absorption performance. Three per cent open area takes the bell-curve shape of a Helmholtz style absorber with a resonant peak around 600 Hz. Zero per cent open area essentially functions as a panel or membrane absorber.

PANEL ABSORBER

Panel absorbers can be designed to efficiently absorb a narrow range of low frequencies. They are often used to in tandem with porous absorbers, which are mostly effective at mid and high-frequencies and must be very thick to absorb at low frequencies. A panel absorber will require significantly less material than a porous material for frequencies below 250 Hz.

Panel absorbers are made from a thin solid sheet with a surface area of 0.5m² or larger and backed with an air space. The induced motion or flexural vibration of the panel, when struck by sound, will absorb some sound energy (converted to heat energy) and this range of absorption can be tuned to a specific bandwidth of frequencies.

When a thin panel is installed with an air gap in front of a reflecting surface, the panel movement is resisted by the air cavity which functions like a spring. The greater the depth of the air cavity, the less stiff the spring. The smaller the air space, the stiffer the spring.

The resonant frequency of a panel can be roughly calculated with the following formula, which considers the mass of the panel and the depth of the air gap behind the panel. The formula provides a very rough approximation of the resonant frequency, which will typically occur below 400 Hz. At this resonant frequency, the absorption will reach its peak and decline rapidly above and below the resonance.

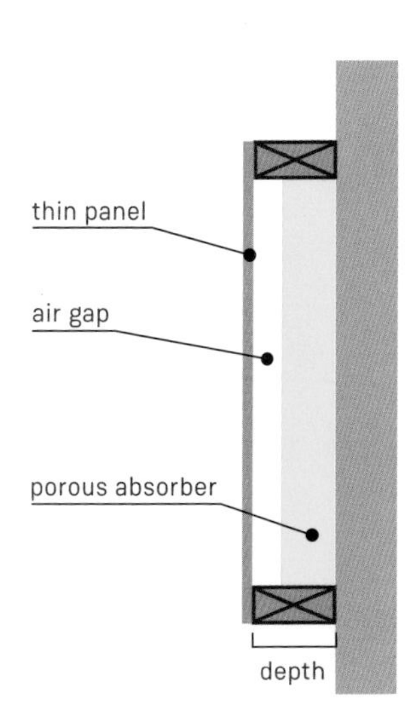

$$f = \frac{x}{\sqrt{m * d}}$$

f = frequency (Hz)
x = constant (60 for metric; 170 for imperial)
m = mass of panel (kg/m^2 or lb/ft^2)
d = air gap distance (metres or inches)

A variety of materials can be used to make a panel absorber such as plywood, plastic, metal, glass, or even paper. The mass of a panel can be calculated by weighing a piece of the material with a known size and thickness. Alternately, there are many online resources that list weights for common building materials.

Using a porous absorber (fibreglass or mineral wool) within the air cavity behind the panel increases the damping of the system and will broaden the width of the resonance and provide absorption over a wider range of frequencies. The porous material should be spaced away from the panel so that it does not limit the movement of the vibrating panel.

Low-frequency panel absorption occurs in common construction where panelised materials are used - such as plywood, glass windows, or drywall partitions. For example, 12 mm thick drywall attached to wood studs is shown to have an absorption coefficient of 0.29 at 125 Hz. The low frequency absorption provided by such assemblies can be beneficial in many applications. However, in spaces for unamplified music performance, this degree of absorption can be detrimental, shortening the reverberation times at low frequencies, which may be desirable for supporting the music.

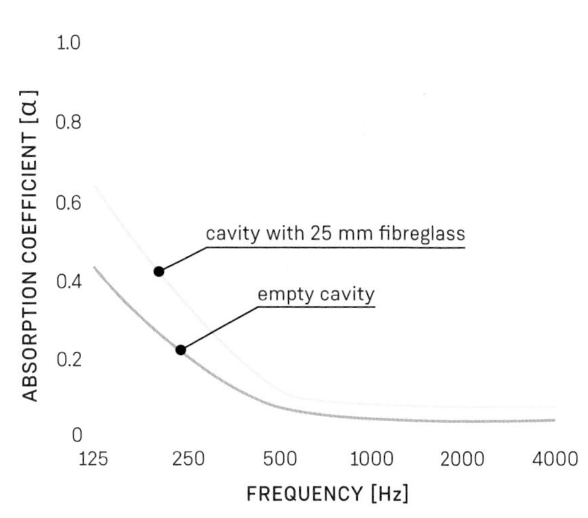

Sound absorption coefficients of a 6 mm thick plywood panel spaced 75 mm from the wall — with and without a 25 mm thick fibreglass blanket in the backing cavity.

ACOUSTICAL MASONRY UNITS

Acoustical masonry units are perhaps the most straight-forward example of a Helmholtz resonator. An opening in the face of the masonry unit functions as the neck of the absorber, while the core of the hollow block serves as the cavity. The bandwidth of resonance is broadened by adding absorptive material in the cavity. In humid or harsh environments, this porous absorber can be encapsulated in polyethelyene or similar film.

The acoustic performance can be designed according to the size and shape of the openings in the block or by changing the volume of the interior cavity to suit the needs of different environments and applications. Flat-faced blocks can potentially create issues with specular reflections. This is resolved with blocks that have angled or diffusive faces to scatter reflections.

By nature of their construction, standard concrete masonry units are porous and moderately absorptive at mid and high frequencies. However, once painted, the porosity is sealed and the absorption is reduced. Acoustical masonry can be painted and still provide resonant sound absorption.

Acoustical masonry units can be structural and load-bearing, with the same compressive strength as standard hollow units of similar composition. Installation is similar to working with conventional masonry units with little additional labour. Due to the heavy construction, these units may also provide sound isolation. Acoustical masonry units are commonly used in industrial settings, outdoor environments such as highway soundwalls, mechanical equipment rooms, and gymnasia.

A single acoustical concrete masonry unit (type RSC) by The Proudfoot Company and produced by ORCO. The front slot cavities are backed with black fibreglass blankets. Two large straight-through cavities allow for applications requiring vertical reinforcement, thermal insulation, or pathways for vertical conduits or piping. When painted and with fibreglass blankets in the cavities, the block achieves NRC 0.85.

Carrie Busey Elementary School, Savoy, Illinois, designed by CannonDesign (2012). The school is home to the district's Deaf and Hard of Hearing programs. The building features a variety of design elements suited to the needs of hearing impaired students. The indoor gymnasium doubles as an assembly space and cafeteria. To control reverberation, the upper walls have been constructed of acoustical masonry units and the exposed ceiling is perforated acoustical metal deck.

The MIT Chapel, Cambridge, Massachusetts, designed by Eero Saarinen (1955). The non-denominational chapel was dedicated as an island of serenity, a space for contemplation and interfaith worship separated from the academic campus. The building forms a 10 metre brick cylinder, encircled by a shallow moat and topped with an aluminium bell tower. Natural light beams from a skylight and glitters down a metal sculpture designed by Harry Bertoia. The curved walls feature locations of open masonry, which are backed with porous blankets to provide sound absorption at the height of the seated congregation.

Songpa Micro-Housing, Seoul, Korea, designed by SsD (2016). This residential tower is comprised of 14 'unit blocks' that allow residents to either claim a single space or combine their block with their neighbours for larger configurations. The ground floor and basement level feature communal spaces accessible to the public. The micro-auditorium, pictured, doubles as a communal living space and cafe that residents and the public can share during the day. In the evening, the space can transform into a theatre, with wedge-shaped terraced seating. The walls and ceiling of the micro-auditorium are clad with acrylic DeAmp panels to control reverberation while allowing natural light and views to the exteriors.

DEAMP

DeAmp is a brand of slotted absorbers that provide sound absorption through the viscous losses as air passes through the small slots. The panels are installed with an air gap and the inherent damping of the system eliminates the need for fibreglass or other porous materials in the air cavity between the panel and the reflective surface behind. The sound absorption performance will vary according to the mounting depth or air gap distance. Panels may be installed with dual layers to provide sound absorption over a wider range of frequencies.

DeAmp panels are available in steel, aluminium, acrylic PMMA, or PETG plastic and in thicknesses ranging from 2-20 mm. The panels can be fastened in a variety of fashions such as display mounts, stand off mounts, or wire-hung. The slit pattern can be manufactured as linear, triangular, sinusoidal, or custom .

SLATS/GRILLES

Slats, or grilles, are most commonly made of wood strips but can also be made from metal or other materials. The grille will be acoustically transparent when there is sufficient open area between the slats. As the gap between slats decreases, or the slat thickness or slat depth increases, the system will begin to function like a Helmholtz absorber with a peak resonance frequency and an upper and lower bandwidth of absorption. Slats may be installed on the walls or ceiling and can be fabricated with a variety of profile sizes, dimensions, shapes, curvatures, and material finishes.

Black acoustical board is often used directly behind the slats to provide sound absorption. The black scrim facing does not reflect light, so the material is not visible. Alternately, coloured fabrics can be used to provide visual accent in the grille gaps while concealing the absorptive backing material.

Slats create visual screening of the ceiling plenum while allowing for airflow from mechanical systems. For ceiling installations, slats may be attached to a suspended grid or rail system. A wide variety of manufacturers provide prefabricated panelised systems with slats held in place by dowels, wood, or metal backing strips.

Assemble is a Melbourne-based residential property developer focused on small footprint projects and publisher of Assemble Papers — an online and biannual print journal about the culture of living closer together. For their Northcote studio, the firm designed and constructed their own acoustical slat ceiling to balance the exposed concrete and glass finishes.
The ceiling design grew out of a process of experimentation, folding different sculpted origami forms in paper. The final construction uses standard sized pine studs and battens harvested from a sustainable plantation. The design utilises five triangular forms, mirrored and repeated five times across the length of the ceiling, evoking the inverse form of a gabled roof.

The National Assembly Building for Wales (Senedd), Cardiff, Wales designed by Richard Rogers Partnership (2005). The undulating Western Red Cedar wood slat ceiling carries seamlessly from the exterior to the interior, forming a funnel at the core of the building that serves as a natural light well for the lower level assembly hall. The ceiling system was custom manufactured by BCL Timber Projects using a panelised layout of only 35 different panel shapes, 99 per cent of which could be flat-packed and curved on site without the use of steam-bending. A mineral wool backing provides acoustic absorption throughout the entire interior.

Samples of Armstrong Metalworks perforated sheet metal panels in a variety of painted and simulated wood finishes.

PERFORATED SHEET METAL

Sheet metal is one of the most common materials to perforate for covering porous absorption and there are a vast number of products and manufacturers. Perforated sheet metal is practical because it is very thin, strong, durable, and can be bent and formed into a variety of shapes. In addition to architectural environments, perforated sheet metal is used to protect porous material in a number of industrial noise control applications such as lining the walls of machinery enclosures, HVAC duct silencers, and acoustic louvres. In harsh or sensitive environments, the porous material can be loosely encapsulated in a thin film behind the perforated facing to further protect the absorber from dirt and moisture.

Perforated sheet metal is available in a variety of hole shapes, patterns, dimensions, gauges, and types of material. With advancements in CNC fabrication, customised shapes and perforation styles are becoming easier, quicker, and cheaper to produce. When designing a sheet metal panel for acoustical transparency, it is generally recommended to have a perforation open area of 20 per cent or more. Perforated sheet metal can be finished to create a variety of visual appearances — it can be painted, coated, digitally printed upon, or finished with veneers or films to appear like other materials such as wood. At a distance, small perforations may not even be visible to the eye.

WOOD PANELS

Wood is a popular choice for fabricating transondent facings or creating Helmholtz style absorbers, particularly for interior finishes. Wood panels can be drilled or routed in a variety of ways to provide the necessary open area access to a porous backing material. Wood is generally thicker than sheet metal and, due to the increased panel thickness, some designs may require larger size or greater number of perforations to achieve absorption performance that is comparable to sheet metal.

Wood veneers are often applied to the top of more cost-effective and sturdy substrates such as medium-density fibreboard (MDF). It is common for this substrate to be broadly perforated with a very high open area percentage. This allows the layer of thin wood veneer to have a smaller percentage of open area and still function as an effective Helmholtz absorber.

Left: Samples of Armstrong Woodworks perforated panels with an assortment of veneers, from left to right: Maple, Light Cherry, Dark Cherry, Walnut Espresso, Cherry, Beech, Walnut, and Wood Wheat.

Right: Armstrong Woodworks Channeled panel, 25 mm thick with 3 mm grooved openings. Dark cherry veneer is applied to the surface of a MDF core board with large circular perforations.

5

Hamad International Airport Passenger Terminal Complex, Doha, Qatar, designed by HOK (2014) The passenger terminal evokes a sense of flow through the architecture's form and surface. The massive, undulating ceiling unifies the interior spaces and creates an evolving texture of pattern and light as one progresses through the terminal. This ceiling is comprised of thousands of acoustically absorptive perforated metal tiles. HOK teamed with CeilingsPlus, a leading manufacturer of metal ceilings, to develop a 3D model of the ceiling surface so that each tile could be custom designed to produce the complex curvature and integrate with architectural elements such as skylights. The tiles were fabricated at a rapid rate of 54 seconds per panel, covering more than 400,000 square meters. In addition to the main concourse, acoustical metal ceilings were used in prayer rooms, boarding areas, first class lounges, and fixed jet bridges. In total, there were over 20 different kinds of metal acoustical ceiling types produced by CeilingsPlus with a variety of aesthetic finishes that included different kinds of woods such as eucalyptus.

ACOUSTICAL METAL DECK

Metal deck is a versatile building material that can be used for the construction of lightweight roofs or as a composite system for creating poured concrete floors and ceilings. In lightweight roofing applications, metal deck can be flat, pitched, or vaulted into complex forms.

Metal deck is often left exposed as an interior finish. There are a variety of options available in terms of profile dimensions and aesthetic appearance. Metal decks are commonly available with an acoustic option by incorporating a porous absorber behind perforated metal.

With a corrugated deck profile, absorptive material typically fills the flutes, which are perforated along the sides. Cellular type metal decks have flat perforated faces and the absorption fills the cell cavity. Perforated flutes tend to perform less efficiently than the flat surface of perforated cellular deck, particularly at high frequencies. This is because at certain angles of incidence, the sound waves do not enter the flute perforations and are reflected.

The absorption performance of acoustical decks typically ranges between NRC 0.5-0.9. One of the primary benefits of using acoustical metal decking is that acoustic absorption is uniformly integrated directly into the building structure with a single product. Furthermore, the absorption is hidden from view and protected from damage by the perforated metal facing.

Assorted metal deck profile options from the manufacturer Epic Metals, from top to bottom: Toris CA, Toris 5.5A, Wideck WPA, Wideck WHFA, Super Wideck.

Goodes Hall, Queen's University School of Business, Kingston, Ontario, designed by The Ventin Group (2013). The project expanded upon an existing historic Richardsonian Romanesque school house from the late 1800s and introduced a new atrium along the heritage building's east side. The atrium features an amphitheatre style design with terraced seating that overlooks the University commons. A lightweight roof, composed of an acoustical metal deck (Super Wideck by Epic Metals), controls reverberation and supports a variety of functions such as lectures, presentations, public congregation, or quiet study.

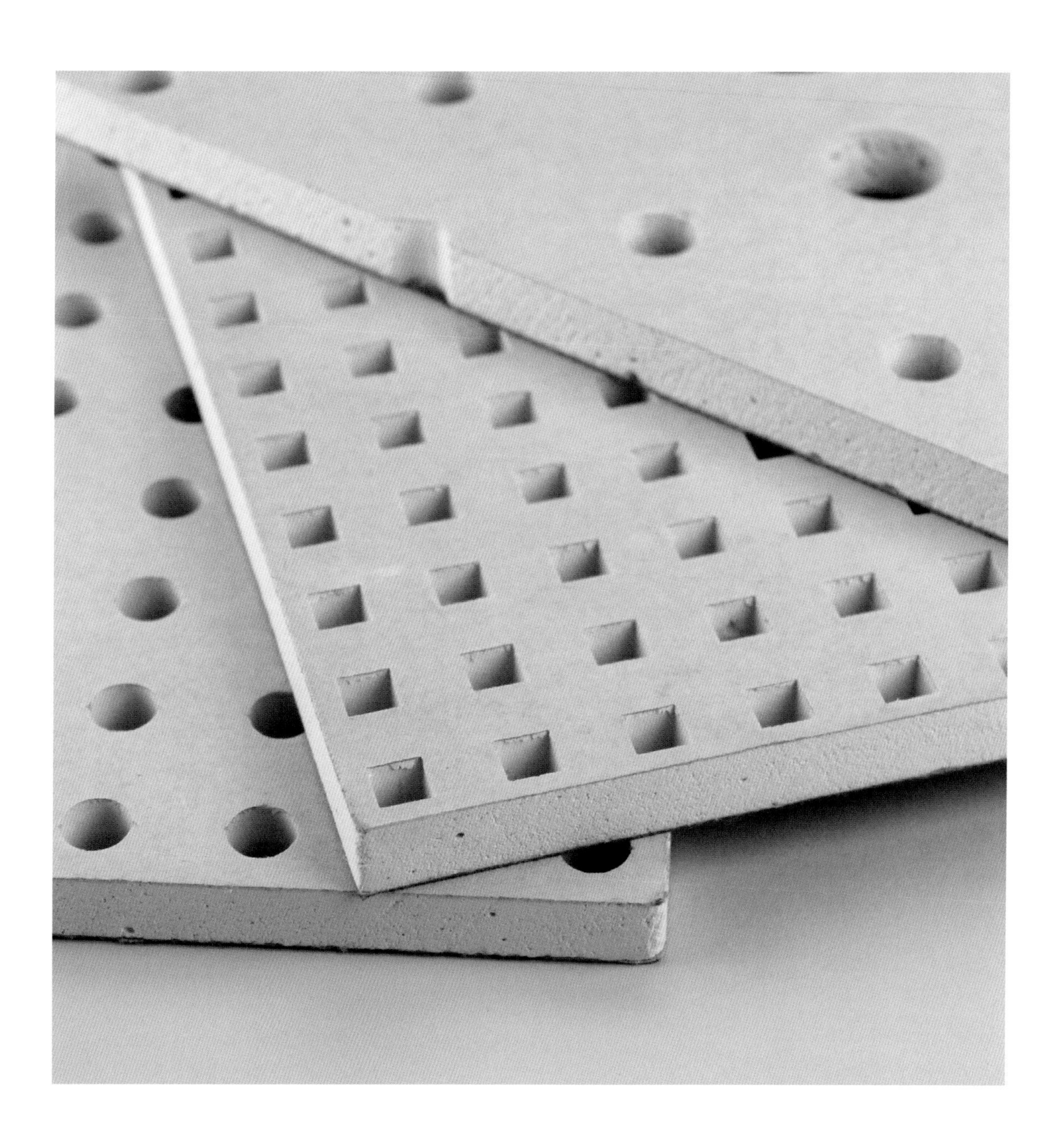

PERFORATED DRYWALL

Gypsum board or drywall is composed of gypsum plaster pressed between thick sheets of paper and commonly used as the outer layer of walls and ceilings in standard building construction. When gypsum is perforated, it can function as an acoustic absorber. An endless supply of perforation patterns are available — circular, square, slotted — in various combinations and densities. Most of these perforated gypsum materials have an open area of 20 per cent or less, which means they tend to function as Helmholtz or resonant absorbers, requiring an air gap and some acoustically absorptive backing material such as mineral wool to achieve broadband absorption. Most products have a non-woven fleece backing that prevents any fibrous backing material from being visible or poking through the perforations.

Perforated gypsum installs similarly to regular drywall, mechanically fastened to furring channels or a suspended framing system. The joints between boards may be sealed to create a monolithic surface and boards can be fabricated to accommodate curved or vaulted surfaces. Typical NRC values range from 0.4-0.9 depending upon the perforation size, perforation density, thickness of material, backing material, and cavity depth.

In some parts of the world where gypsum is less commonly used, perforated wall board is made of materials such as magnesium-oxide or calcium silicate, which function acoustically in the same manner.

One of the benefits of perforated drywall is that it can be painted with a roller or brush without any impact to the sound absorption performance. Once installed, paint should not be applied with a sprayer as it may clog the perforations or coat the porous backing material and reduce the sound absorbing performance.

The University of Pécs Faculty of Music and Visual Arts, Pécs, Hungary, designed by GEON Architects (2010). Drywall with varying perforation patterns is used throughout the interior walls and ceilings to create a dynamic, playful, yet quiet environment for music practice rooms, performance spaces, and classrooms.

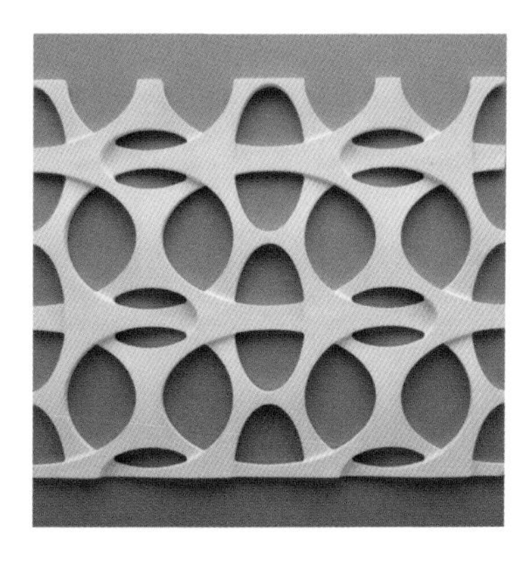

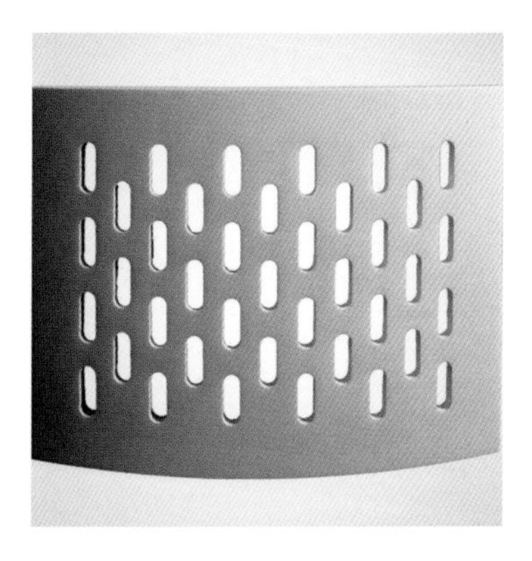

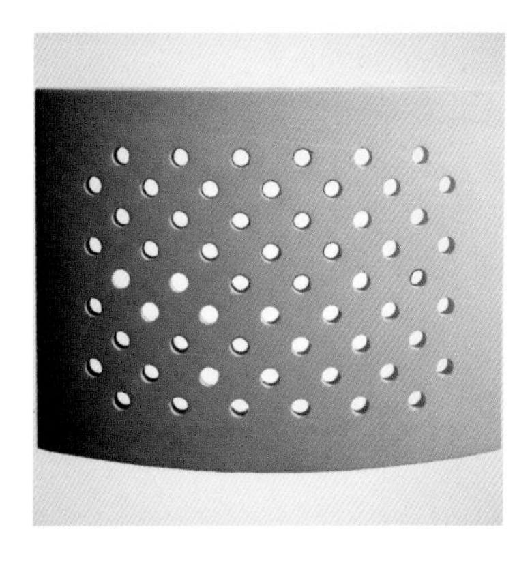

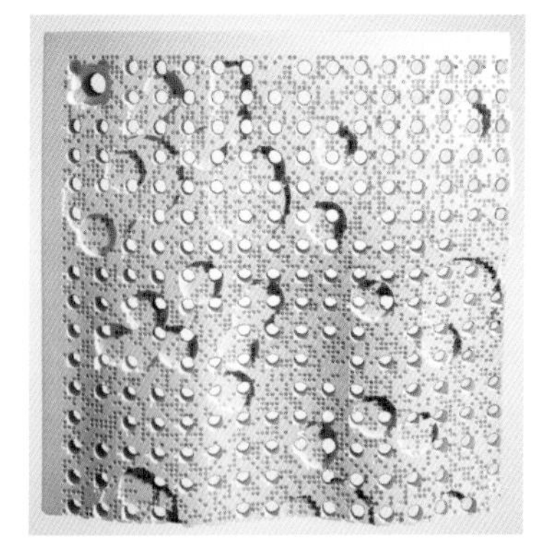

FORMGLAS PERFORATED GFRG

Formglas is a leading manufacturer of Glass Fibre Reinforced Gypsum (GFRG) cast products. The material can be perforated or cast to be acoustically transparent or function like a Helmholtz absorber, similar in nature to perforated drywall or perforated wood.

Glass Fibre Reinforced Gypsum is a composite of high strength alpha gypsum reinforced with glass fibres that can be factory moulded into virtually any shape or size. GFRG is a desirable alternative to traditional plaster castings due to its light weight (10-15 kg/m^2), strength, ease of installation, and use of post-consumer recycled material. The material is non-combustible and may be factory finished or, more commonly, field finished with any interior paint.

Formglas GFRG is highly customisable and offers limitless design possibilities in terms of casting forms, patterns, textures, and finishes. 5-axis CNC machining allows custom perforation patterns and shapes and the application of perforations to curved surfaces.

GFRG is typically cast with a 5 mm thick shell and enhanced edges that are 19 mm thick. The panels may incorporate embedded steel or wood elements for added strength or to provide hardware for attachment or suspension. Panels may be face-fastened with screws through built-in reinforcements that are counter-sunk and filled. For a monolithic finish, panels are manufactured with tapered edges so that joints may be taped and finished similar to gypsum wall board.

Samples of Formglas perforations; the style numbers for the samples are listed from top to bottom: 50167; 50165; 50164; 50163.

JFK International Airport, Delta Airlines Terminal 4, New York, New York, designed by Skidmore, Owings & Merrill (2013). The sinusoidal, vaulting ceiling of the concourse is composed of perforated GFRG Formglas panels.

TOPAKUSTIK

Topakustik is a line of wood acoustical panels produced by NH Akustik + Design in Switzerland. The products are primarily comprised of a perforated medium-density fibre (MDF) core board which supports a thin finished layer of wood veneer, melamine, or high performance laminate. Each panel typically has a fleece backing and is intended to be installed with an air gap and a porous backing material, such as glass or stone wool. Panels can be installed on walls, suspended ceilings, and even integrated into cabinetry and doors. The manufacturer provides numerous details for creating clean edges, corners, cut outs, and borders.

The standard Topakustik finishes are linear grooved (with narrow, medium, and wide spacing options) or perforated (with large, medium, small, and micro-perforated options). The perforation size or the width of the groove will result in different visual appearances — as a faint texture for small perforations or as a distinct visual pattern for larger perforations.

Acoustic performance can be tailored to the needs of each product according to the core board perforation size, the finished layer groove or perforation size, and the cavity or mounting depth behind the panel. Topakustik offers an 'M' type core perforation pattern that is designed for mid-to-high frequency absorption, and a 'T' type perforation pattern that is designed for the low-to-mid frequency range.

A variety of customisable options are available, such as curved panels, custom veneers, and with the CNC capabilities, custom perforation patterns.

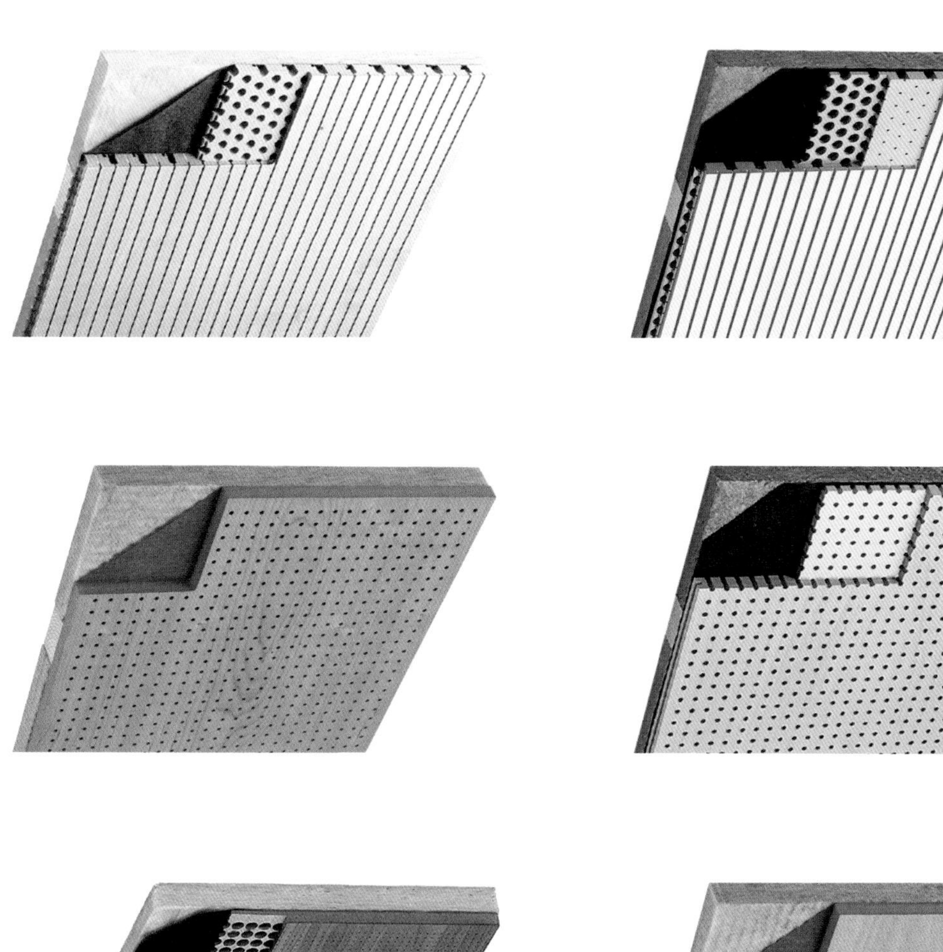

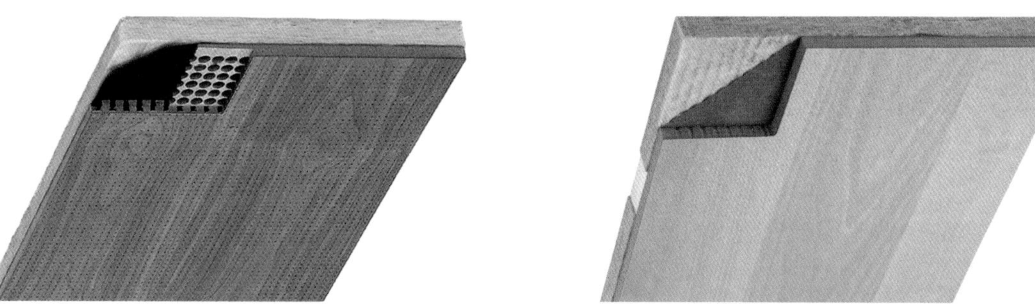

Samples of Topakustik products. Left column, top to bottom: Topakustik Type 14/2 M; Topperfo Type T; Topperfo Type Clou. Right column, top to bottom: Topakustik Type 13/3 M; Topperfo Type M; Topperfo Type Micro.

Franco-German Brigade Dining Hall, Immendingen, Germany, designed by hotz + architekten (2011). The dining hall provides meals for 1500 soldiers and officers each day. With an acoustically reflective ceiling and floor, sound absorption is provided along the walls with a combination of Topakustik type 9/2 M and Topperfo type M 8/8/5, which are finished with Canadian maple veneer and semi-gloss varnish.

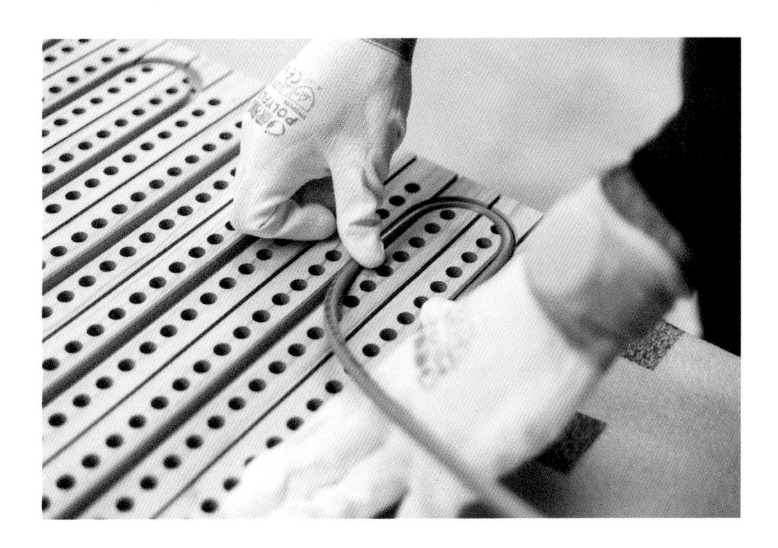

CLIMACUSTIC

Climacustic is a modular panelised system that combines radiant thermal heating or cooling with sound absorption into a single product. The system uses a 35 mm thick EPS or mineral wool insulating layer bonded to aluminium sheet metal which is attached to the wall or ceiling. The outer acoustical panels are composed of MDF, finished with melamine or veneer, which join directly to the aluminium sheet metal. Sandwiched between the sheet metal and MDF panel is a PeRT tubing fluid circuit that provides the radiant heating or cooling. The modularity of the panel system allows for easy access and repairs. The system can be installed as a flat or curved surface.

Climacustic is produced by Fantoni, a family owned company founded in 1882 which grew into an industrial furniture manufacturer in the 1920s. The company earned a permanent place in the history of furniture making with the introduction of the Serie 45 range designed by architect Gino Valle, which was added to the permanent collection of the Museum of Modern Art in New York in 1974. In addition to furniture, the company manufactures a number of milled and drilled MDF-based sound absorbing panelised systems, some of which can be integrated directly into their furniture, desk partitions, cabinetry, or doors.

M Power Yoga, Baltimore, Maryland, designed by J Neal Architecture (2015). This hot yoga studio features the FantoniUSA Climacustic ceiling which provides radiant heating and sound absorption to accommodate music and instruction during classes.

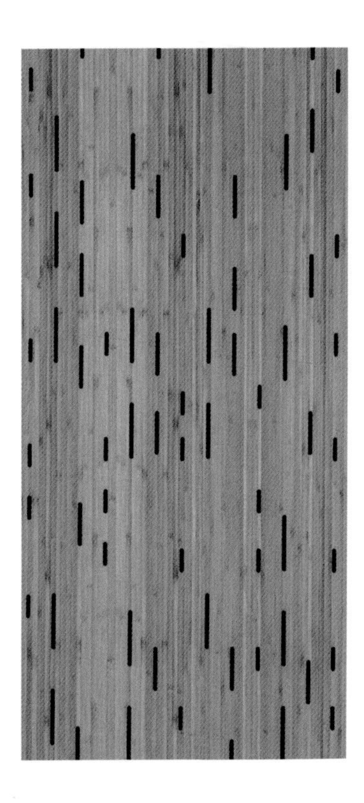

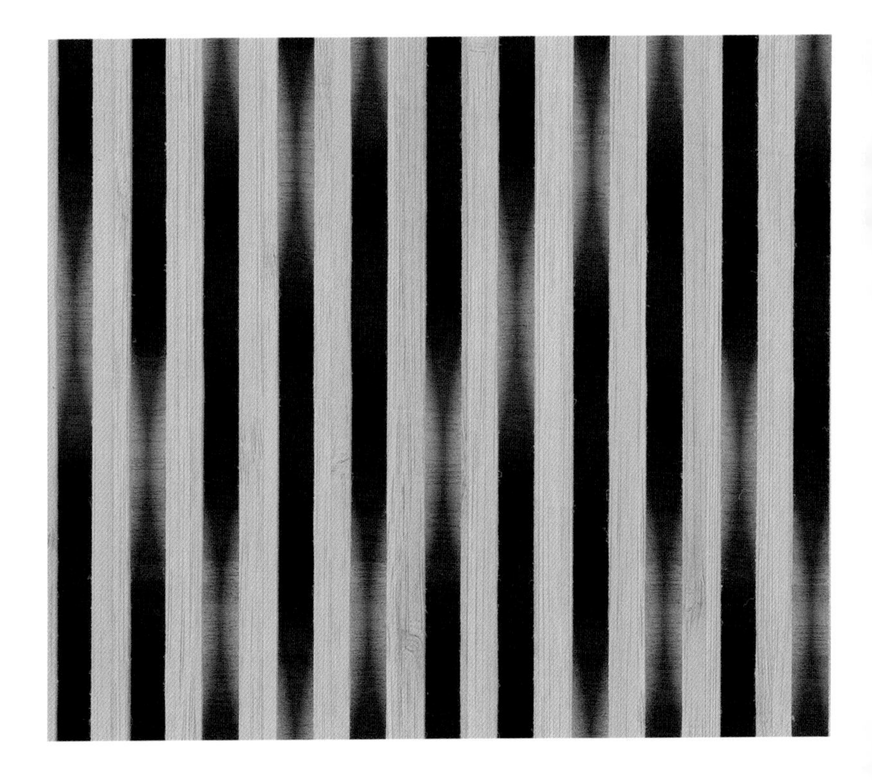

PlybooSound panel style A2 (far left) and LinearSound style LS14 in fog colour.

PLYBOO

The Plyboo brand, by Smith & Fong, introduced bamboo flooring and bamboo plywood to the United States in the 1990s. In recent years, the brand has expanded into acoustical panelling with the PlybooSound and LinearSound product lines which offer a variety of perforation patterns, grain patterns, finishes, and textures.

Plyboo products are made from fully matured Moso species of bamboo, which is rapidly renewable and certified by the Forest Stewardship Council. The Moso species grows to maturity in five to six years to a height of over 12 metres and a diameter of more than 15 cm. Each year, only 20 per cent of the plantation's bamboo — only the five-year growth — is cut, ensuring the forest canopy remains intact and the ecosystem is not disturbed.

Standard panel sizes are 1219 mm x 2438 mm x 19 mm thick (4 ft by 8 ft by 0.75 inch thick) and are available factory finished or unfinished. Panels can be installed on walls or ceilings. Test reports have shown the panels achieve NRC ratings between 0.5-0.7 when mounted directly against a wall with a 25 mm absorptive backing material.

Left: An installation of PlybooSound panel style A3.

Below: PlybooSound panel style A5.

DUKTA

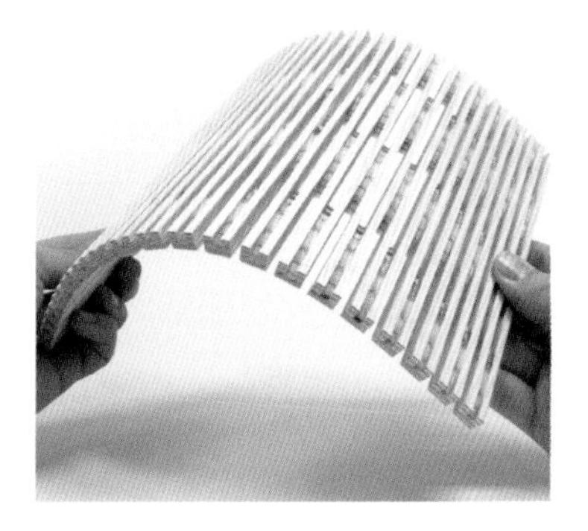

Dukta is manufactured with a patented process of machined incisions on one or both sides of wood (or wood-based) material to provide pliability and transparency. The resulting open area typically ranges between 10-40 per cent depending upon the particular incision pattern. With a large percentage of open area, Dukta may function as an acoustically transparent screen, whereas with smaller open area it may function as a Helmholtz style absorber. The product is typically backed with mineral wool or similar porous material to provide broadband sound absorption. Dukta's 'Janus-Tex' features an intermediate felt layer that provides some integrated absorption when the material is installed as a baffle or room divider.

The concept for Dukta emerged from a student design course assignment to construct a lounge chair using curved wooden elements. The company founders, Serge Lunin and Christian Kuhn, further developed the Dukta product with support of an 18-month CTI (Commission for Technology and Innovation) research grant in partnership with the Institute for Design and Technology at the ZHdK; the architecture, wood and construction departments at Bern University; and Schreinerei Schneider AG in Pratteln.

In 2011, the company was founded to bring Dukta to the marketplace. Dukta has since been used on large projects such as the Toni-Areal concert hall and cinema and two large Wood Loop exhibitions in the Winterthur Museum of Commerce and the Brengenzerwald Exhibition Centre in Andelsbuch.

Dukta is continuing to develop new products and work with new partners in production and distribution domestically and internationally. Additional applications for Dukta include furniture, partitions/screens, and luminaires.

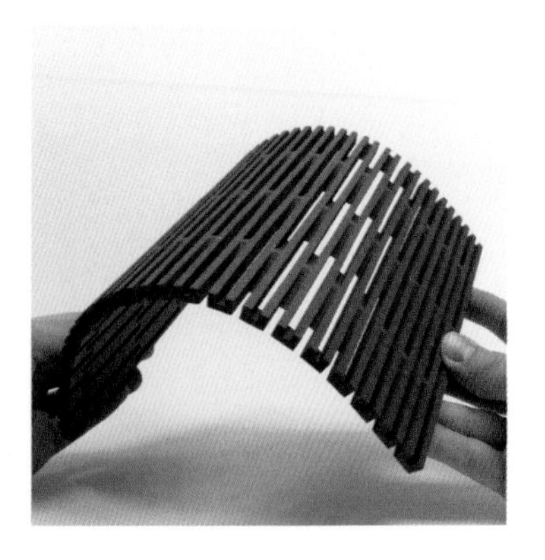

Left: Dukta samples, from top to bottom: LINAR in birch plywood; JANUS TEX in spruce with blue fabric core; FOLI in MDF; SONAR in black MDF.

Opposite: The Toni-Areal concert hall at the Zurich University of the Arts (ZHdK), designed by EM2N (2014). The concert hall features dukta-LINAR with acoustically absorptive backing along the lower half of the rear and side walls.

ORGANOID

Organoid Technologies is an Austrian company that creates acoustic panels with decorative coatings made of raw materials applied to high pressure laminate support structures. One of the original decorative coatings is made from hand-cut alpine hay which is grown 1700 metres above sea level at the base of the Wildspitze Mountain in Austria. The hay is combined with daisies, roses, lavender, or vanilla, which gives the panels a distinctive aroma. Decorative coatings are made from a variety of other materials such as bark, flowers, herbs and fruit.

The coatings are combined with various binders and, for fire safety, a flame-retardant phenolic resin is used. The coatings are acoustically transparent, allowing the core structure to provide sound absorption, which may be a slotted or perforated resonant type absorber, or backed with wool for porous absorption.

The company prides itself on the use of sustainable production practices such as using green electricity and binders that are 100 per cent biodegradable and free of biocides, plasticisers, and solvents.

Organoid decorative coatings are designed for vertical light-duty indoor applications or roofed outdoor applications and should be protected from prolonged moisture and humidity to prevent premature aging, swelling, shrinking and the decrease of the natural aromas of the decorative coatings.

The perforated Picture Absorber with recessed cavity to hold porous absorber and a linear Helmhloltz style absorber.

A variety of decorative coatings.

FLEX ACOUSTICS

Flex Acoustics creates variable absorbers specifically to control low and mid-frequency reverberation times in large performing arts venues. Longer reverberation times at low frequencies may be desirable for certain types of unamplified music but problematic with amplified rock, pop, and dance music where articulation of the bass and the beat is important. Low-frequency absorption can be difficult to obtain with porous materials, which must be very thick or spaced away from walls.

Flex Acoustics has two patented product lines (aQflex for permanent installation and aQtube for temporary events) which are thin, plastic membranes that absorb sound when inflated with air. Test data indicates these membrane absorbers have a peak absorption around 500 Hz. The shape and dimensions of the material when inflated provides diffusion of sound above 1 kHz. Once deflated, the system stops absorbing sound, allowing for longer reverberation times.

Because the system is essentially plastic and air, it is lightweight and can be easily installed on ceilings or walls with a track or suspended wire. Absorption will depend upon the surface area of coverage and the spacing of the material. When desired, the system can be visually hidden behind an acoustically transparent ceiling such as a grid or scrim.

aQflex installed at a cultural center in Ulsan, Korea.

aQtube installed at an abandoned shipyard, which served as the venue for the 2014 European Song Contest.

Louis Namèche swimming pool, Molenbeek-Saint-Jean, Brussels, designed by Altiplan Architects (2015). Filva-T panels are installed at intervals along the exterior facade glazing to control noise within the natatorium while still allowing natural daylight to filter in.

FILVA-T

Filva-T gets its name from the contraction of the words 'film-vibration-transparency.' The material is comprised of a thin, clear, fluoroethylene film held between two sheets of aluminium metal mesh. Sound absorption occurs due to the vibration of the film membrane. The material is typically installed with some air gap or freely suspended because an air space behind the panel is a necessary component of the absorption mechanism.

Filva-T is also shown to block sound and has been used in barrier wall construction for highways. The panels are only 1.5 mm thick, lightweight (1.4 kg/m^2), UV-resistant, washable, and weather resistant. Test reports indicate peak sound absorption reaches up to 0.9, with a resonant frequency that is dependent upon the distance of the air space behind the panel.

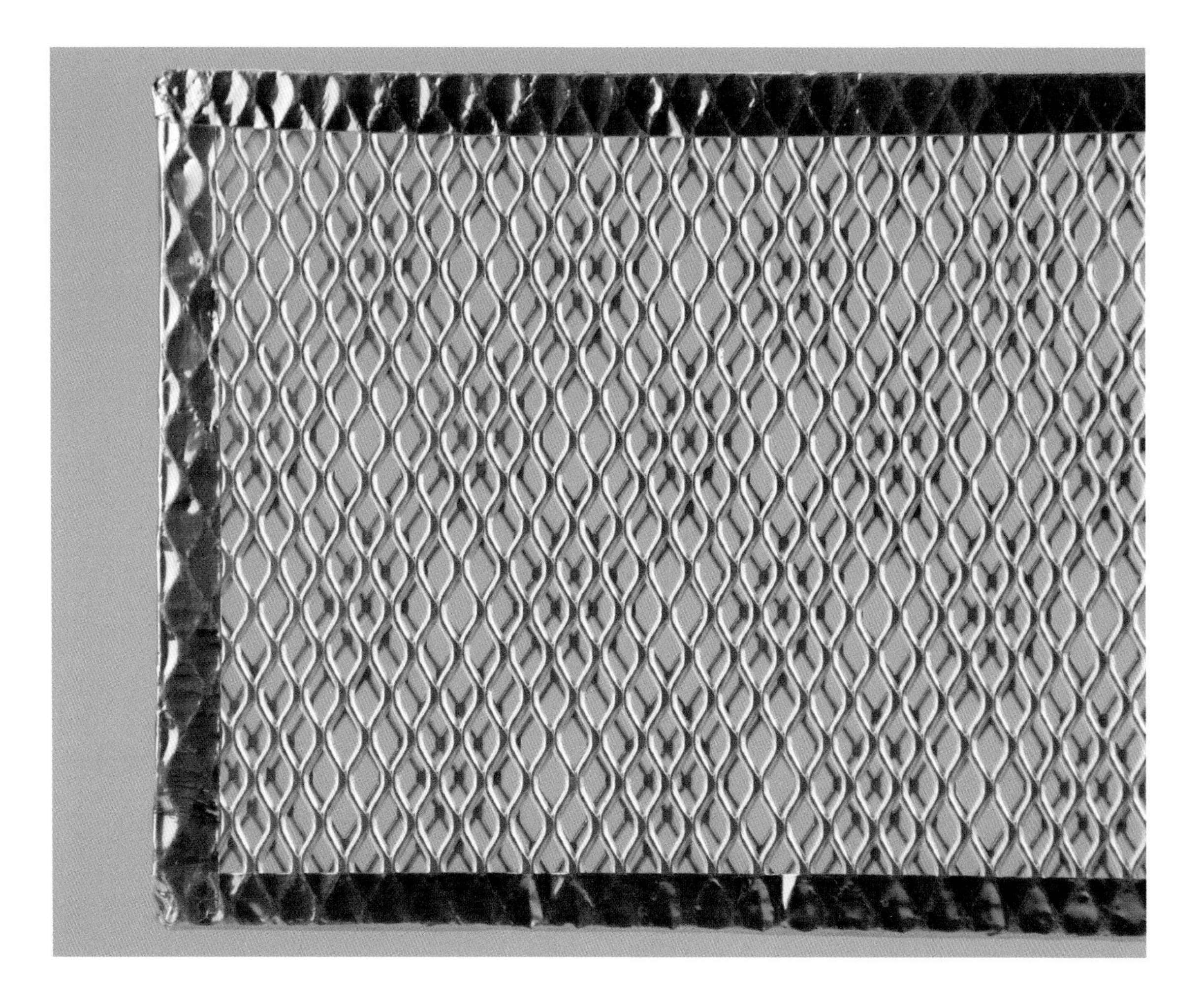

Above: Textile Softwall illuminated internally with LEDs.

Opposite Above: Softwall in yellow textile, kraft paper, and blue textile.

Opposite Below: Cloud Softlight Mobile, internally lit by LED.

MOLO SOFTWALL

Softwall and Softblock, by molo, are free-standing, pleated partition systems designed for subdividing spaces. The vertical pleats provide stiffness and allow the material to expand, contract, and curve. The product is constructed from either kraft paper or a non-woven polyethylene textile, commonly known as Tyvek. The Tyvek material offers a paper-like appearance that is translucent, can be dyed, and is tear, UV, water, and flame resistant. Kraft paper and Tyvek are not porous materials, so sound absorption occurs due to diaphragmatic movement of the many thin-walled cells created by the honeycomb pleats.

The Softwall system is modular and units can be attached and held together with small magnets. A number of accessory components have been designed by molo to integrate into the partitions such as LED lighting and seating. Acoustic tests of the softwall have achieved NRC ratings between 0.45-0.60. The pleated Tyvek designs have also been extended to lights, mobiles, and pendants, called Softlight, which have also been tested for sound absorption.

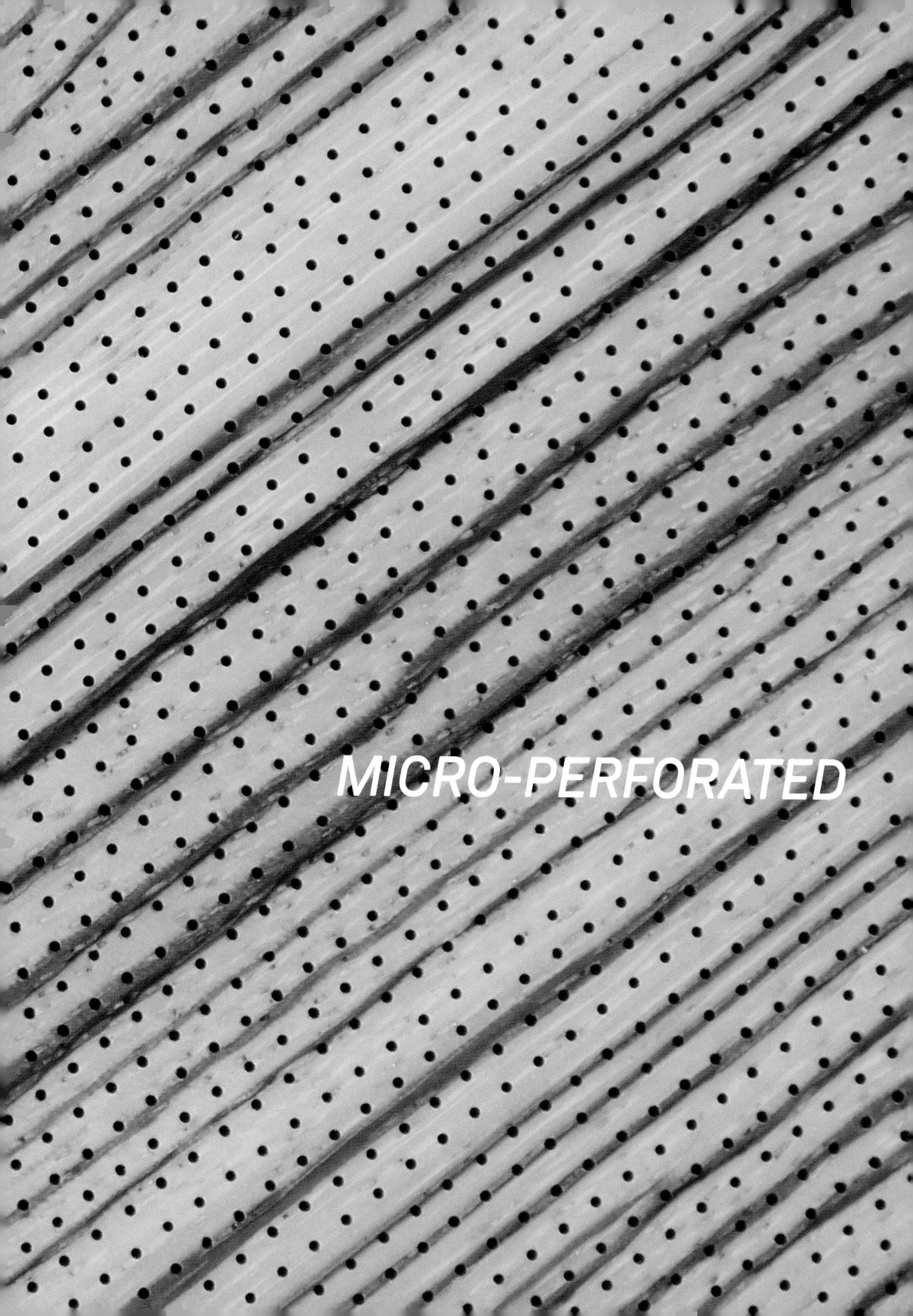
MICRO-PERFORATED

OVERVIEW

Micro-perforated absorbers are one of the newest sound absorption technologies to emerge. The theory was first introduced in the 1970s by Daa-You Maa and commercially available materials have rapidly developed since the 1990s.

As discussed in preceding sections, rigid materials can be perforated to provide a transparent screen for porous absorptive materials (when sufficient open surface area is provided) or these panels can function as Helmholtz or resonant type absorbers. As a Helmholtz resonator, a narrow frequency of absorption occurs depending upon the perforation size, the density of perforations, and the distance between the perforated panel and boundary surface. The width of the frequency range of this resonance can be broadened by using a porous absorber behind the panel.

Micro-perforated panels (MPP) function somewhat similarly to traditional perforated panels. However, MPPs have perforation diameters between 0.1-1.0 mm, and a broad range of absorption can be achievable without using a porous backing material.

The small size of the micro-perforations creates enough acoustic resistance for sound absorption to occur. Typically the material being perforated is very thin and dimensionally comparable to the size of the perforations. An endless variety of materials can be used such as plastic, paper, wood, or metal and the fundamental principle of absorption performance is largely independent of the type of material being perforated. Because there is no porous material necessarily required, MPPs may be used for applications where porous absorption is not permitted such as in healthcare, food preparation, micro-electronics fabrication, or clean room environments. Due to their lightweight, MPP materials are desirable for aircraft and automobile applications.

When MPP theory was first introduced, the ability to precisely fabricate hundreds of tiny perforations per square centimetre was a considerable challenge. A number of advanced technologies have emerged in recent years that have made micro-perforation increasingly cost-effective to both prototype and manufacture on larger scales, such as hot and cold needle perforators, punch, laser drilling, and CNC drilling.

The acoustic performance of micro-perforation can be approximated with mathematical formulas which consider the material thickness, perforation size, perforation density, and mounting distance of the panel to the reflecting surface. By adjusting these parameters, a material can be custom designed to achieve the desired absorptivity over a range of frequencies. Like any resonant absorber, there is typically a peak absorption with a bell curve roll-off above and below the peak. The resonance frequency of this peak can be shifted by changing the depth if the air cavity or mounting distance.

MPP absorption performance may be enhanced by using a porous absorber within the backing cavity. As well, it is possible to achieve absorption over a broader range of frequencies simply by using two or more layers of MPPs spaced away from one another. Since panels have different mounting distances, the peak absorption of each panel occurs at different frequencies. The absorption of the layered panels combines to cover a wider range of frequencies.

MPPs are generally very thin materials and they may lack mechanical strength or be susceptible to damage. The use of a honeycomb backing can create a very strong and lightweight composite material. When a micro-perforated sheet is applied to both sides of the honeycomb, it will function like a double-leaf MPP. The use of honeycomb has been shown to cause an increase the peak absorption at the resonance frequencies. The thickness of the honeycomb can be changed to alter the distance between the sheets, which will influence the peak resonance frequencies.

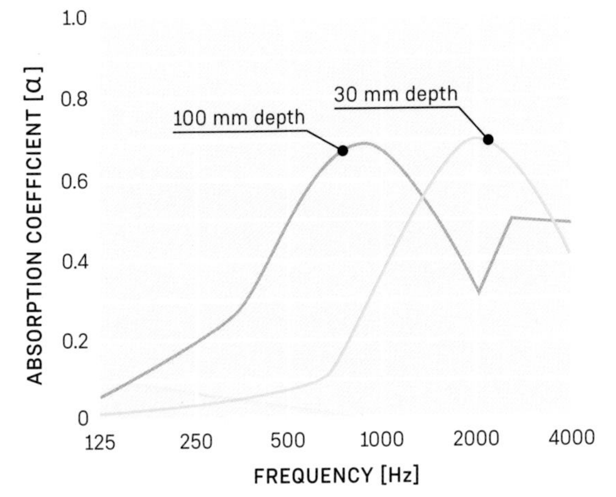

Absorption coefficients of the same micro-perforated foil with two different mounting depths. The foil is 1 mm thick with 0.2 mm diameter perforations.

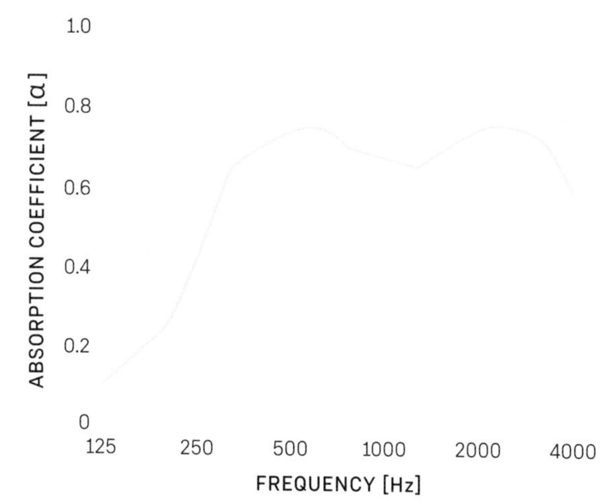

Absorption coefficients of a double-layer of micro-perforated foils. The outer layer foil is mounted at a distance of 100 mm; the inner layer foil is mounted at a distance of 30 mm. Both foils are 1 mm thick with 0.2 mm diameter perforations.

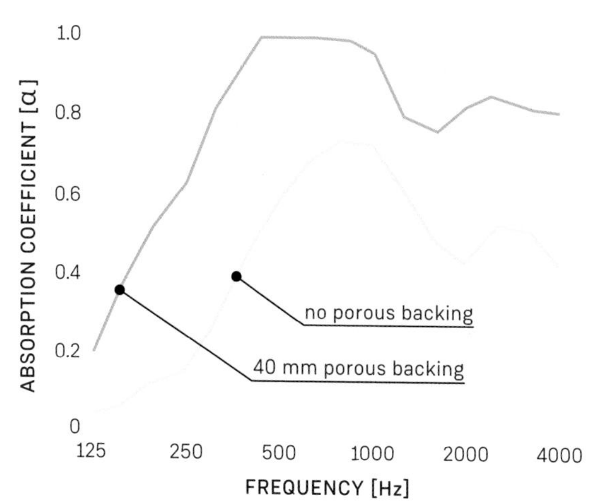

Absorption coefficients of micro-perforated stretched PVC with and without fibreglass absorber in the backing cavity. The micro-perforated material is 0.18 mm thick with 0.15 mm diameter perforations.

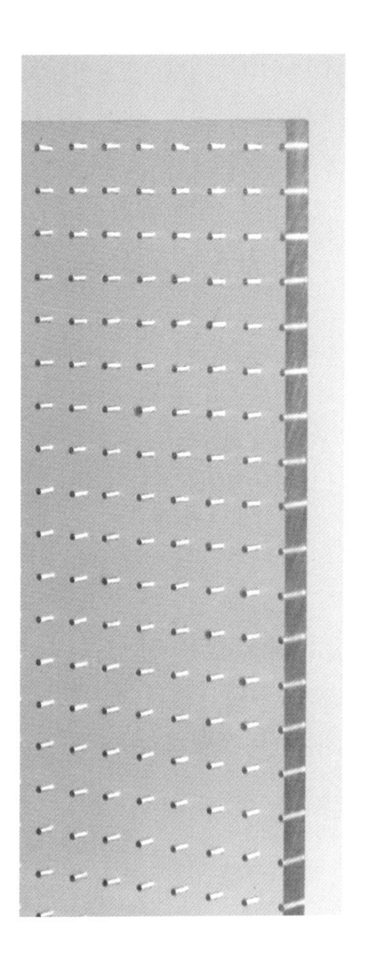

Left: 5 mm thick micro-perforated acrylic panel.

Right: Clearsorber micro-perforated 1 mm thick semi-rigid transparent sheet and 0.3 mm thick flexible foils, transparent and with reflective metallic print.

MICRO-PERFORATED FOILS & SHEETS

Micro-perforated foils are available in transparent, translucent, coloured, and printed foils, most commonly marketed under the names Clearsorber or Microsorber. These films are made of EFTE Teflon or Polycarbonate and less than 1 mm thick. Due to their transparency or translucency, the material can be used to reduce sound reflections from glass without significantly reducing natural daylight. As well, due to the transparency, the material can be suspended within a room without impacting the perceived dimensions. Foils can be integrated into a variety of installation methods but are typically mounted under tension using a hembar or grommets held taut by springs, or draped in a lapendary fashion. Thicker, transparent micro-perforated materials are available in rigid sheet form, made from Polycarbonate, Acrylic, or PETG in thicknesses ranging from 2-15 mm.

The Schluterhof courtyard of the Old Armory (Zeughaus) at the Deutsches Historisches Museum, Berlin, Germany. The glass and steel canopy was designed by I.M. Pei in 2003 to enclose the courtyard and allow the space to be used for events. However, the new glass ceiling resulted in long reverberation times and its curvature created undesirable focusing and echoes. In order to maintain the transparency of the canopy, the ceiling was retrofitted with two layers of micro-perforated foils held together by a steel wire mesh. The corners of each 1.15 m x 1.15 m foil are attached to the nodes of the wire mesh with spacers to maintain a distance of 15 cm between the foil layers. The distance between the foils and the concave glass roof ranges between 7 metres at the center of the canopy to 4 metres at the edges.

STRETCHED MICRO-PERFORATED PVC

Vinyl, otherwise known as polyvinyl chloride (PVC), is a thermoplastic material that can be reprocessed using heat. PVC is non-flammable and can be used to create stretched ceilings that are installed somewhat similar to a stretched fabric system using perimeter tracks or frames with a harpoon clip to hold the material in place. Once the vinyl is stretched into place, it can be further tightened or shaped through the application of a heat source. It is possible to un-stretch and re-stretch the material when heat is applied. Vinyl is available in a variety of colours and finishes which can be micro-perforated to provide sound absorption. Translucent PVC is often backlit to create diffuse lighting effects. The leading providers of stretched micro-perforated PVC systems are NewMat and Barrisol.

Opposite: The St. Louis Zoo Penguin Puffin House, Missouri, designed by Peckham Guyton Albers & Viets Architects (2003). White matte micro-perforated PVC by NewMat creates a large barrel vault ceiling used for theatrical lighting effects to simulate colourful sunrises and sunsets. The PVC material is well suited to the humid environment. The micro-perforations control noise levels for both the animals and the patrons.

Samples of micro-perforated vinyl membranes by NewMat. White matte (ACM15), left, and white lacquer (ACL101), right.

The Modern at The Museum of Modern Art (MoMA), New York, New York, designed by Bentel & Bentel (2004). The 330-seat ground floor restaurant is located at the nexus between the original museum building designed by Edward Stone, the adjacent museum annex designed by Philip Johnson and the newer museum facility designed by Taniguchi. NewMat white lacquer micro-perforated PVC membranes were incorporated into the design of the ceilings for the dining and bar areas, using a custom perimeter extrusion which allows for the ceilings to appear like floating sheets of white glass with a razor sharp edge. The membranes are backed with fibreglass to increase the acoustical performance.

Double-leaf micro-perforated Clearsorber with polycarbonate honeycomb core, 19 mm thick.

DOUBLE LEAF MICRO-PERFORATED HONEYCOMB

Translucent honeycomb panels are commonly composed of micro-perforated PETG sheets held together with a polycarbonate honeycomb core. The composite structure results in a very strong, lightweight, and rigid structure that is translucent or semi-transparent. These materials are available from manufacturers such as Design Composite, Panelite, RossoAcoustic, and Akustik & Innovation. When surface mounted, the panels should be provided with an air gap to provide the best absorptive performance. The panels may also be used for desk partitions, room dividers, sound absorbing privacy screens, and backlit ceilings. Some manufacturers offer customisability in terms of honeycomb core structure design, surface coatings, and printable finishes.

Honeycomb panels are also made entirely from aluminium, one of the lightest and stiffest building materials with a high resistance to moisture and corrosion. As a micro-perforated absorber without any porous or fibrous material, the aluminium honeycomb is finding many applications in transportation, marine, and industrial applications.

CP30 honeycomb panel by RossoAcoustic is used for room and desk partitions to provide privacy, light transmission, and sound absorption in one material.

SonoPerf, developed by Akustik & Innovation, is a micro-slotted metal plate with slot widths of less than 0.4 mm. The metal is provided as powder-coated steel, stainless steel, or anodized aluminium, with custom colour and printing options.

MICRO-PUNCHED METAL

Sheet metal is a good candidate for micro-perforation because it is thin, rigid, and once perforated it can be formed and worked with just like any other sheet metal material. These products are often used for hygienic applications such as healthcare, industrial, or food preparation environments, where porous materials may not be permitted. Micro-perforated metal is also used for engine enclosures and interior cabinets for machinery and appliances such as washing machines. The panels, when mounted with an air gap, provide moderate sound absorption at mid and high frequencies without the use of any porous material. The absorptivity can be enhanced by using a porous backing material or an acoustical fleece backing such as SoundTex.

The Silk Metal Panel from Acoustical Surfaces uses a patented side-rolling grind method instead of direct 90-degree punching to produce micro holes. The steel sheet is 0.38 mm thick and can be finished with anodized white, black, or custom colour printing.

SmartPerf with white and grey micro-perforated veneer.

SMARTPERF

SmartPerf is a new material designed from a collaboration between Akustik & Innovation from Switzerland and Sandler AG from Germany. The panels utilise a flame-resistant non-woven polyester core created with a patent-pending process that makes a low density skeleton of polyester fibres that is highly porous, lightweight, strong, and self-supporting. A micro-perforated outer layer is applied to the core, which can be solid colour, photographic print, or real wood veneer. The panels can be cut to size and may be easily integrated into partition walls or furniture such as cabinets or sideboards. The SmartPerf core does not support bacterial grown, is not damaged by humidity and dries quickly when exposed to water, suggesting this material may find future applications in harsh environments.

SmartPerf lightweight polyester core in white and grey.

MICRO-PERFORATED WOOD VENEER

Wood veneers are a popular application for micro-perforation. When viewed from a distance, the tiny perforations are not noticeable and the interior finish may appear to be monolithic wood. Micro-perforated veneers are sometimes applied as the exterior finish of a resonant Helmholtz absorber, adhered to a thick layer of MDF or similar material which has a larger perforation pattern. In such an application, the panel assembly is mounted with an air gap and often with a porous backing material.

Wood veneers can also be applied to sheet metal, which is then micro-perforated. The composite of the veneer with the sheet metal provides greater stability while maintaining a thin surface, which is desirable for good absorption performance. More recently, micro-perforated veneers have been applied directly to dense fibreglass or fibreglass honeycomb cores, such as Navy Island's SoundPly material. The composite yields a lightweight, strong, and stable material with excellent broadband absorption. Since the fibreglass is integrated into the interior of the product, there is no need for additional porous backing materials.

SoundPly panel by Navy Island with a quarter-sliced Maple veneer. The composite of micro-perforated veneer with absorptive fibreglass core yields high NRC ratings.

ACOUSTIC-LIGHTBOARD

Acoustic-Lightboard, by Richter Akustik & Design, is constructed from a thin micro-perforated facing that is bonded to a honeycomb structure. The back of the honeycomb is bonded with a non-woven acoustic fleece and another layer of facing which is perforated with a larger open area. Sound absorption is provided by the micro-perforation in tandem with the acoustic fleece backing. No additional porous backing material is necessary.

Acoustic-Lightboard can be perforated in a number of standard shapes and patterns. The surface finish layer can be high pressure laminate, wood veneer, custom lacquered finish or raw high-density fibre board that is sanded and finished. The panels are lightweight, dimensionally stable, and have a high fire resistance. Standard panel thicknesses are 16 mm and 19 mm; custom thicknesses can be fabricated.

Deutsches Theater München, Munich, Germany, renovation designed by Doranth Post Architekten (2014). The 1500 seat theatre hosts a variety of theatrical performances, musicals, operettas, and contemporary music. Acoustic-Lightboard was used for more than 450 m² of interior surfaces.

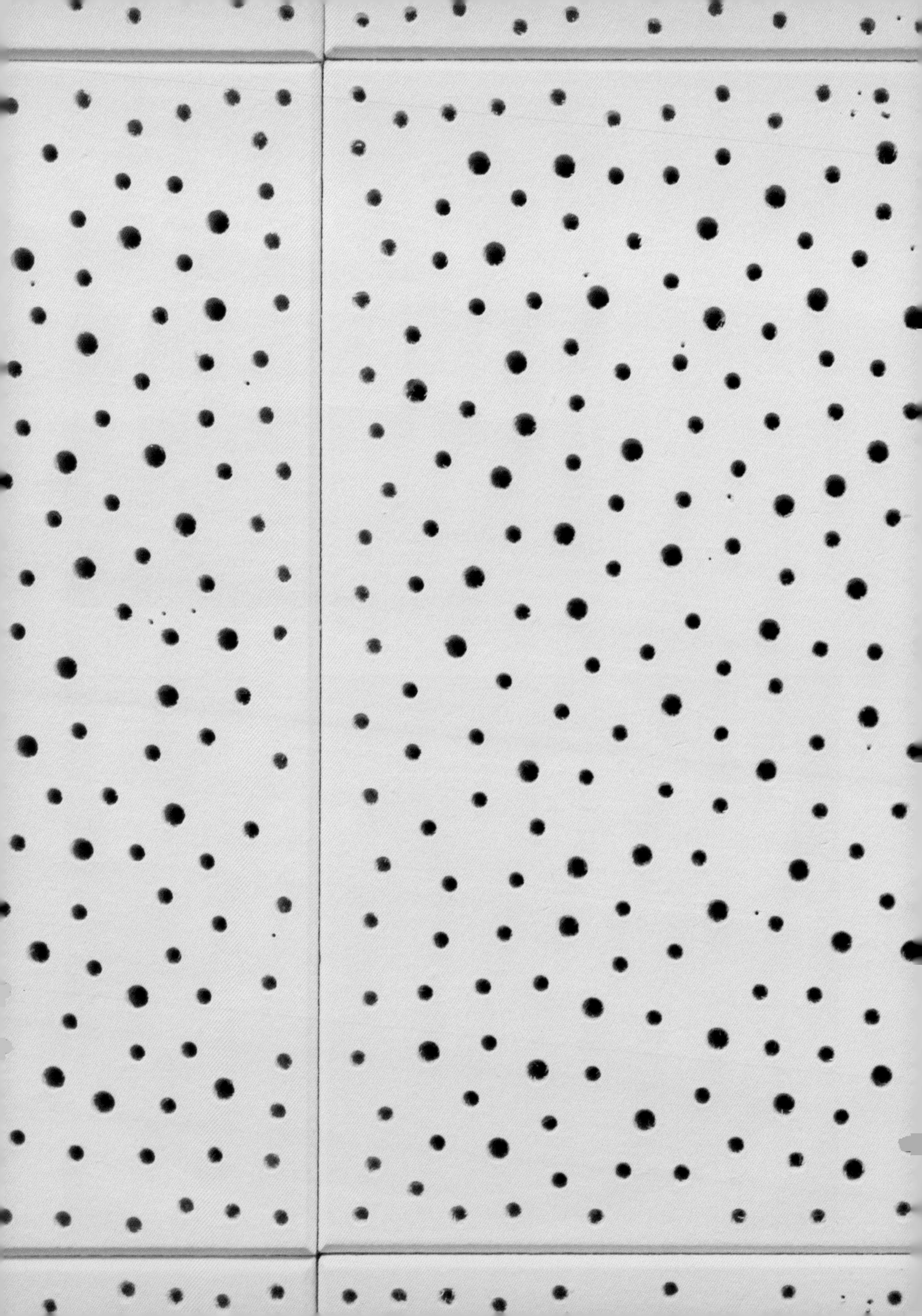

HISTORICAL

Saint Thomas Church, New York, New York, designed by Cram, Goodhue & Ferguson (1914). While the visual design draws upon the great Gothic cathedrals of Europe, the acoustics of the interior are uniquely modern. Rumford Tile provides the appearance of stone blocks that make up the vaulted ceiling but their porous structure was engineered to absorb sound and thereby improve the intelligibility of sermons. Saint Thomas Church is one of the first projects where an acoustical material had been specially designed and integrated into the aesthetic program of a building. In recent years, some of the porous tiles have been sealed to make them acoustically reflective to create longer reverberation times.

OVERVIEW

The history of modern architectural acoustics is said to begin with Wallace Clement Sabine who, in 1895, was tasked with studying and correcting the problematic acoustics of a lecture hall at Harvard's Fogg Art Museum. Speech in the hall was unintelligible due to excessive reverberation and echoes. At that time, there was no quantitative method for evaluating the acoustics of a space because sound recording and reproduction technologies were in their infancy. In his early experiments, Sabine would manually observe the time it took for sounds to decay, often by producing tone from an organ pipe.

While studying the Fogg lecture hall, Sabine would borrow removable seat cushions from a nearby theatre and add them to the lecture hall, noting how the decay of sound decreased in relation to the quantity of cushions. This discovery formed the basis of Sabine's reverberation time equation, which is still used to this day.

As the science of architectural acoustics developed at the beginning of the 20th century, acoustical materials specifically designed to quiet the noise of the boisterous modern era also began to emerge. By the 1930s a plethora sound absorbing materials were in use, made from a vast variety of materials that included sugarcane, flax, jute, licorice, asbestos, eelgrass, cork, wood fibre, vermiculite, pumice, gypsum, lime, and volcanic silica.

One of the earliest engineered acoustical materials was developed in 1911 out of a collaboration between Wallace Sabine and Raphael Guastavino, a Spanish architect and builder who is well known for the Guastavino Tile vaulting techniques, which created self-supporting arches and timbrel vaults using interlocking terracotta tiles and layers of mortar. These vaulted spaces, once finished with hard ceramic tile, resulted in notoriously long reverberation times.

Sabine and Guastavino worked to design a tile which had a high level of surface porosity. The first result of their collaboration was the Rumford tile — a mixture of 25 per cent clay, 10 per cent feldspar, 65 per cent peat. During the ceramic firing process the peat burned away, leaving behind a network of tiny pores which absorbed 29 per cent of sound at 500 Hz. The first project to use Rumford tile was the gothic-style Saint Thomas Church (1914) located at Fifth Avenue and 53rd Street in New York City.

Sabine and Guastavino expanded upon their collaboration to further develop the 'Akoustolith' product line which launched in 1916 and was an aggregate of pumice bonded with Portland cement. Akoustolith had improved absorption compared to Rumsford and was offered as a tile or plaster that could be finished in a variety of styles or cast into custom reliefs.

A number of acoustical plasters followed in the wake of Akoustolith, typically composed of gypsum mixed with fibrous or porous aggregate material that would create the porosity necessary for sound absorption. A few plasters, such as Hushkote, created porosity through an off-gassing chemical reaction that occurred once the plaster was mixed with water. Once the plaster dried and the gas escaped, small pores remained.

Felt is one of the earliest materials to be employed for acoustic absorption. Felt was somteimes installed using furring strips to provide an air gap, which would enhance the sound absorption performance. To protect or obscure the felt from view, it was common to cover the felt with a permeable membrane such as cloth or canvas. By using very coarse mesh cloth, the surface could be decorated with paint without significantly reducing the ability for sound to be absorbed into the felt backing.

Opposite: Advertisement for Akoustilith from *Sweet's Architectural Catalogue,* 1931.

Types of Materials Made by Guastavino

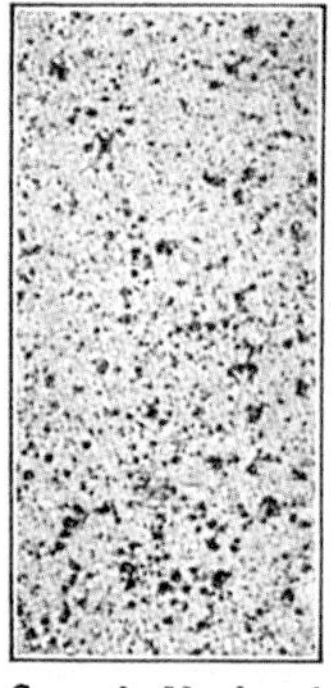

Smooth Unglazed Ceramic Tile

Sections of Ceiling Ribs Cast in "Akoustolith"

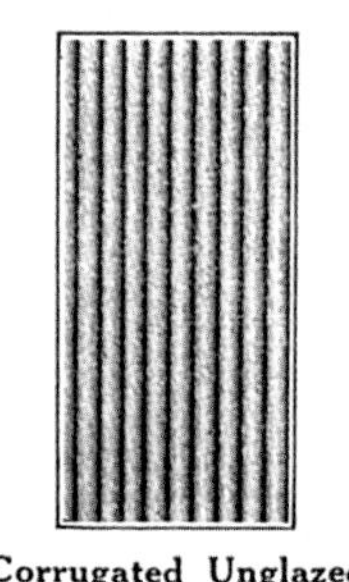

Corrugated Unglazed Ceramic Tile

Decorative Glazed Tile Inserts

Ornament Cast in Our Acoustic Casting Plaster

Decorative Glazed Tile Inserts

"Akoustolith" Tile (Broken to Show Structure)

"Akoustolith" Plaster

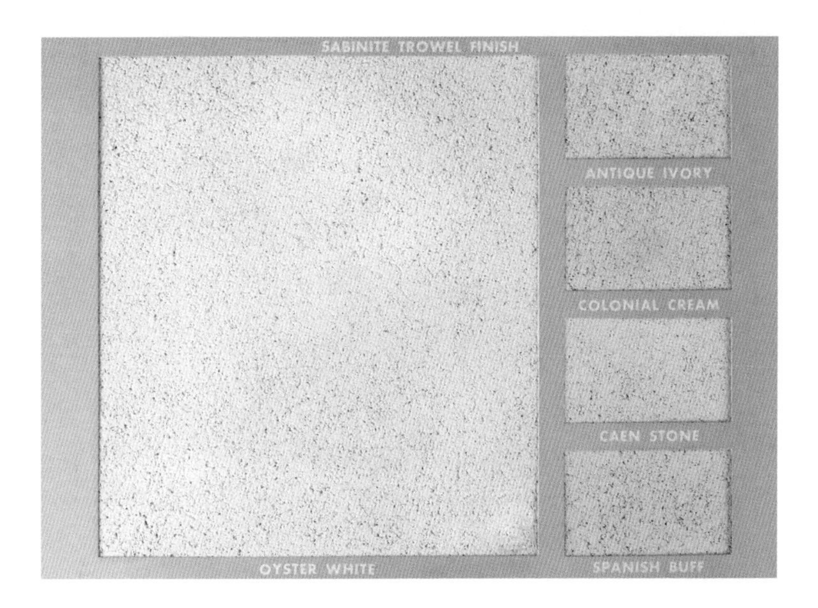

Sample box of Sabinite gypsum based plaster in standard colours, by United States Gypsum.

Johns-Manville created a number of acoustical felts such as 'Nashkote,' which was a canvas laminated to thick Akoustikos Asbestos Felt that came in perforated, solid canvas, or oilcloth finishes. These felt/membrane type systems were susceptible to damage, difficult to maintain, and were custom designed and installed for each project, requiring more installation time than standardised tile or board products.

Board and tile products were made from a wide variety of both natural and man-made materials. The use of standardised modular units enhanced the speed and efficiency of installation compared to felt or plaster products. These materials were often directly adhered to walls or ceilings or mounted over furring strips.

Some of the early tile and board products were marketed for dual thermal and acoustical performance. Celotex, for example, began as a thermal insulating and building board but was observed to have noticeable sound deadening properties. By adding perforations to the surface of the Celotex material, the surface could be painted or decorated and still allow sound to penetrate to the absorptive core or fibrous backing material. Acousti-Celotex, which launched around 1927, was made from bagasse, the fibre waste of sugar cane after the sugar had been extracted. It was manufactured by cooking and washing the cane fibres to remove soluble material. The fibres were then put through a patented 'ferox process' that coated the material with a chemical for waterproofing and to prohibit rot, termites, and vermin. The fibres were then felted and interwoven, without the use of adhesives, into continuous boards that were then heated and dried.

A number of board and tile products gradually fell out of production beginning in the 1950s with the rise in popularity of the T-bar suspended ceiling, which continues to be widely used to the present day. The T-bar system remains popular because of its low cost, modularity, ease of installation, standardised sizing, accessibility, ability to conceal unsightly building services, and the ability to easily incorporate lighting, HVAC, sprinklers, and many other elements.

Many of the fundamental techniques and core materials of these historical products remain in use today in some for another, such as gypsum plasters, cellulose, wood wool, etc. In fact, a few large companies that began manufacturing the earliest acoustical materials are still in business — Johns-Manville, United States Gypsum, Armstrong, among others.

It should be noted that asbestos fibre was a common ingredient in many early acoustical materials because it was both fibrous and provided the fire resistance that is desirable for interior finishes. Asbestos, while a naturally occurring mineral, poses a significant health risk if the airborne fibres are inhaled. Since the 1970s, asbestos production has been banned or heavily restricted in many countries. In most locales there are strict regulations regarding the removal of building materials containing asbestos, typically requiring licensed personnel with specialised equipment. Appropriate precautions should be taken when handling historic acoustic materials.

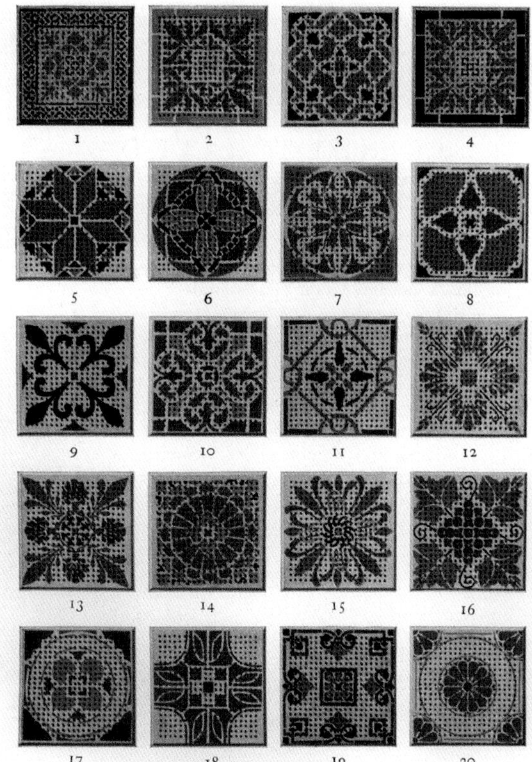

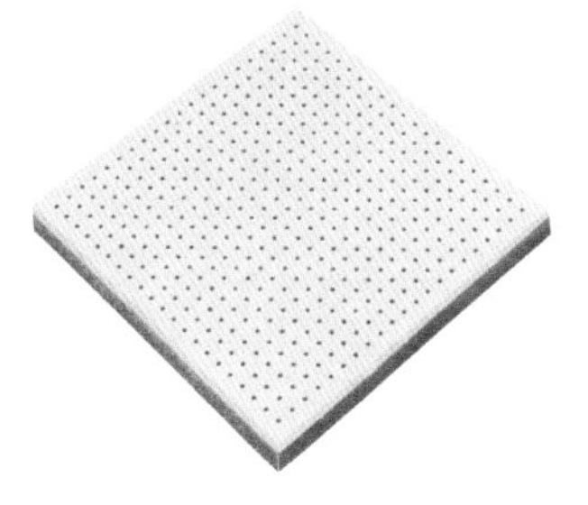

Above: Plain Acousti-Celotex perforated panel, made from cane fibre.

Left: Suggested pre-decorated Acousti-Celotex Tile, from Celotex Corporation's Catalogue entitled *Less Noise Better Hearing*, 1927.

INFORMATION

Union Station, Los Angeles, California, designed by John and Donald B. Parkinson (1939). Considered the last of the great railroad stations built in the United States, the architecture is a mixture of mission revival with art deco and streamline moderne details. The station was placed on the National Register of Historic Places in 1980 and remains a remarkably intact time capsule. Great care was taken in designing the acoustics of the station, which is built primarily of reinforced concrete. Acoustical materials cover nearly every wall and ceiling and are integral to the visual character of the space. On the ceiling, the recessed coffers feature ornate, hand-painted perforated acoustical panels. The walls of the Ticket Lobby and Entry Lobby are clad in Calicel, a material composed of cellular silicate of lime and alumina that is expanded 20 to 40 times its volume, mixed with a mineral bonding agent, moulded in a hydraulic press and then kiln-dried. The walls of the Waiting Room feature a material composed of mineral fibres designed to look like natural stone. The sense of peace and quiet in these spaces is a stark contrast to the raucous bustle of the concrete concourse that leads to the train platforms. Today, Union Station is a major transportation hub, serving over 100,000 passengers each day. The spaces are frequently used for civic meetings, lectures, live music, and performing arts events.

INTERVIEWS

JEFF GREENE

EVERGREENE ARCHITECTURAL ARTS

Jeff Greene is the President and founder of EverGreene Architectural Arts, a specialty contractor and design studio working with commercial, government, institutional, sacred, and theatre clients in the areas of interior restoration, conservation, decoration, and new design. The company operates as both an art studio and a contractor combining art, science, and technology as designers and craftsmen work side-by-side.

EverGreene Architectural Arts is presently working to restore the acoustical finishes within the Ticketing Lobby of the Los Angeles Union Station. What does this process entail?

The scope of work is to clean and replace missing or damaged acoustical wall panels throughout the Ticket Lobby, Foyer, and Waiting Room. Architectural Resources Group did the initial testing and then we were called in. The material in the Ticket Lobby is called Calicel, made

Crews remove damaged and loose Calicel tiles at Union Station. For the tiles that remain in place, the cleaning process brings the Calicel tiles back to their original, lighter colour.

of an expanded stone aggregate, mixed with a mineral bonding agent, moulded in a hydraulic press and then kiln-dried. The cleaning process involves dry-cleaning by gentle vacuuming and brushing to remove the build-up; mild abrasive cleaning; or wet cleaning to maintain the amber patina of the naturally aged tile. We tried about six or eight different methods just to clean the tiles and the results of this are pretty spectacular.

We also needed to replace a number of broken, scarred, patched, and discoloured tiles. Originally it was 219 tiles but as we went through and tapped each tile, we found it was double that number. I can tell you that there have been all kinds of strange things that have happened to this place in terms of maintenance and repairs over the years. There was a section of tiles that was a slightly different colour and what we discovered was, they weren't acoustic panels! Someone had trowelled in a material and then scribed in joints to make it appear like a tile. Nobody was aware of this until we pointed it out.

Another thing we're doing is, where there are surface blemishes, we're able to cut back the face a little bit and expose it and it doesn't make it too bright, and if we need to we can tone it down by staining it to get it all in balance with one another because this place has been retrofitted with all kinds of band-aids.

How do you recreate an historical acoustical material? Is it a trial and error process? Manufacturing processes aren't typically published information and early acoustical materials were made from a wide variety of strange things.

We do a lot of pre-construction work, research, and testing that informs how a project is going to be done. Essentially what happened here, the architect came in and did their archival research and said here's what the material is. We came in to solve how to recreate it. It was really based on our empirical knowledge and familiarity with different materials and we were able to get the sample right within a week.

Detail views of Calicel samples show the tiles, from top to bottom, before cleaning, after cleaning, and a newly fabricated tile to match the original appearance.

When we looked at this particular aggregate material, it had a rounded quality. Perlite usually has a much sharper look to it. We also needed something that would hold all the aggregate together without filling all the holes, something lightweight. So we tried some relatively dry mixes, plasters, glues, a number of different things. Then we would back it up with a secondary material so we would get the base.

When reproducing an acoustical material, are you focused upon the visual appearance? Or is there an effort to match the original acoustical performance?

Generally we're matching the aesthetic appearance. We find that a lot of times the original material has been painted and is no longer acoustically absorbing. Sometimes that's desirable, sometimes it's not desirable.

For instance, the Fox Theatre in Tucson, Arizona has a faux-Travertine acoustical material called Acoustone that was produced with bicarbonate of soda, which off-gassed when it was mixed with water. The off-gassing created tiny pores throughout the entire material that made it acoustically absorptive. When we recreated the material, we used a similar process but only focused on the surface layer. The material had been painted over the years and the acoustic properties had already been modified or compromised, so the client was only concerned with achieving the original appearance.

On the other hand, when we worked on the Mormon Tabernacle, they were insistent that any changes to the material would not alter the acoustic characteristics of the space. We worked with an acoustical engineer and acoustical measurements were taken beforehand. We then produced large mock-ups that were installed on a scaffold where sound was bounced off the material, measured, analysed in a computer, and compared with the original material measurements to be certain it would match the original characteristics.

Should the aural dimension of architecture be preserved with the same intentionality as the visual?

Historic preservation is about what people value and what meaning they imbue in objects and buildings. To me the appearance or function may be bricks and mortar but the more important thing is about how people interact and perceive something in preservation. In Europe there seems to be a much broader interpretation of preservation focusing on how a building can be adapted so that it's a living building as opposed to a time capsule.

The thing about acoustics is that acoustic tastes have changed. For example, we've worked on about 450 theatres. Originally, you had theatres with natural acoustics that amplified the sound from the stage and bounced the sound out into the auditorium. Now, you have electrically amplified sound and different types of music, so the space needs to function differently.

In a church, you may have different needs for spoken word or singing. We restored a synagogue in Cleveland that had an Akoustolith ceiling. The client wanted to increase the reverberation time and so we applied a sealant to the surface of every other tile to render half of the tiles reflective.

The use of the building evolves and acoustic needs evolve with the use of the building. I think the only time to preserve the acoustic function is when the building is still being used in the way as it was intended when it was built. ⁘

New tiles are mixed, moulded, and then antiqued to match the original Calicel tiles.

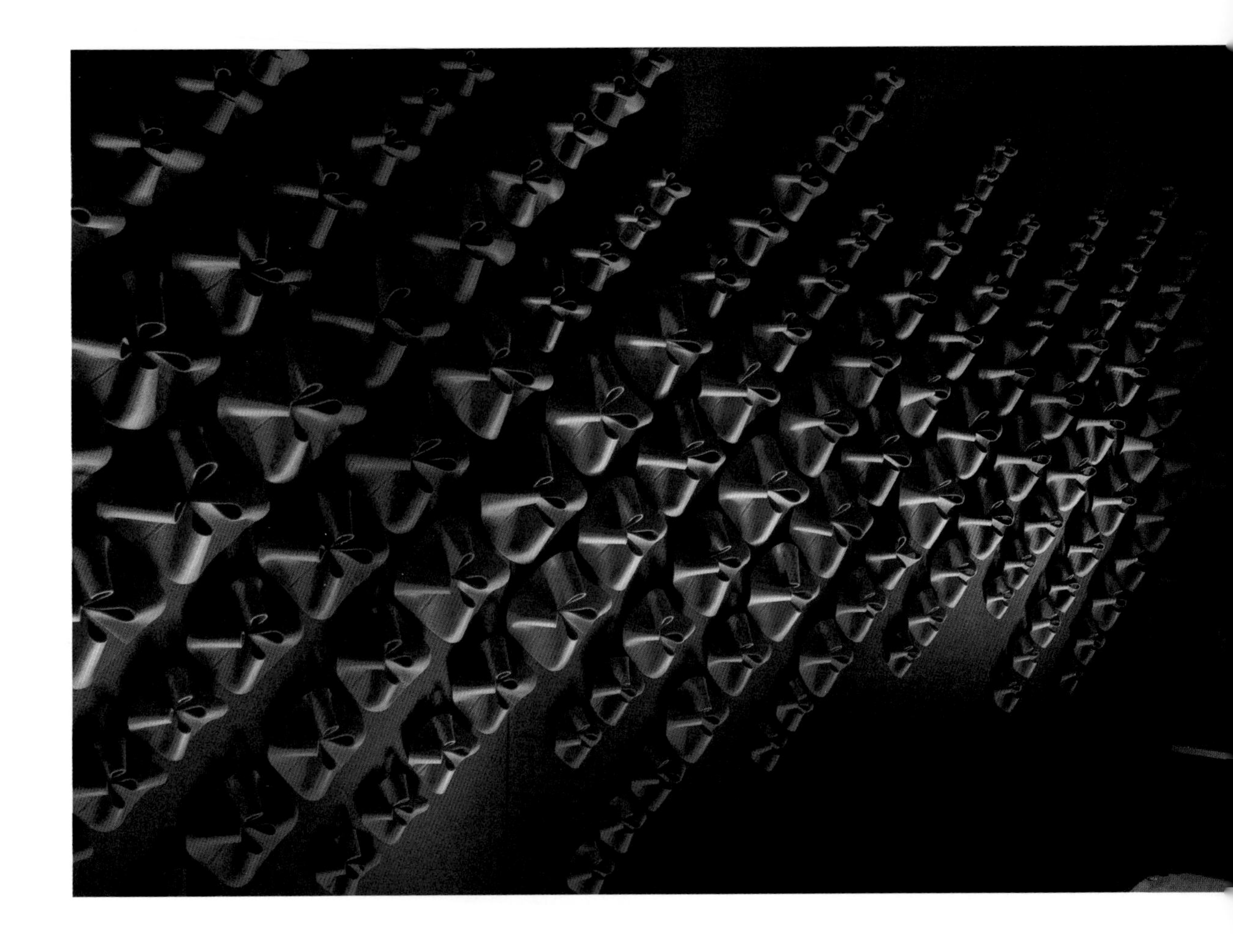

ANNE KYYRÖ QUINN

London based designer Anne Kyyrö Quinn works at the intersection of design, craft, and art, creating custom-made, hand-fabricated, architectural textile wall installations. Her work is known for its highly graphic, three-dimensional construction, bold colours, richly textured surfaces, sculptural silhouettes, and minimalist motifs.

Your background is in textile design. How did you become involved in creating acoustical treatments?

While I was a textile design student, I experimented with many different materials including felt – I found the material very inspiring. Subsequently, I started using more and more felt in my designs and settled on 100 per cent wool felt as being the best material to form my sculptural textile structures. I also like felt because it is natural, durable, and versatile, which makes it an ecologically sound choice.

A custom designed wall feature for the auditorium of the 2015 100% Design Fair, London, UK.

After a number of different projects creating decorative textile wall hangings and feature walls, my clients noted the sound-deadening, calming quality in the places where the pieces were installed. So we decided to get various designs tested to find out the sound absorption coefficients of the designs. The designs were thoroughly tested and classified in accordance with International Standards (ISO) 354 and 11654. The tests proved the designs were ideal for spaces where noise control and reduction is a concern. The test results also gave us a good base line to further develop and research techniques to improve the acoustic performance of the 3D textile creations. This led me to further study architectural acoustics at London South Bank University's Acoustics Department.

We started on the basis that the textile structures we designed should look good and function on an emotional level. Most acoustical products, at least at that time,

Opposite Above: Cable Curtain design (2014) for a reception lobby in Paris, France.

Opposite Below: Tulip and Rosette wall textiles at the Sesame Street Headquarters (2011), New York, New York.

were designed to be functional rather than attractive - we strive to achieve both.

All the products are custom-made to individual, often site-specific requirements. Because of the flexibility of the manufacturing process, material, and the customized designs, we can work with technicians and acousticians to achieve the required solution for a given environment. We can also augment the inherent sound absorbing qualities of the designs with other acoustic solutions concealed behind our panels, but this of course would depend upon the room, the type of noise, etc.

Why do you enjoy working with felt?

I use 100 per cent wool felt - the material possesses at the same time the softness of wool and a certain rigidity that allows me to create sculptural 3D designs that are highly tactile and when lit correctly, can create an impression of dynamic movement in a static object. So for me at least it ticks many boxes. Although I am best known for using felt - which was not a fashionable material for a textile designer to be using at the time I started to work with it — I also work with a number of other natural materials such as leather and silk depending upon the customer's requirements. However, the felt forms the structure of the designs and provides the great sound absorbing qualities.

The combination of natural felt and three-dimensional patterning has proven to be the main factor in my felt textiles as sound absorbers. My signature three-dimensional surface technique has been engineered to provide greater surface area and flow resistance, critical in the absorption of sound. This enables the felt wall coverings to significantly reduce ambient noise levels and reverberation in any environment.

How have you integrated acoustical engineering into your design work?

I have worked very closely with London South Bank University's Acoustics Department for several years. My designs get tested by the department where I was a student of Environmental and Architectural Acoustics. The ongoing collaboration with the LSBU Acoustics Department has proven invaluable in the development of the acoustic performance of my work. The head of the Acoustics Department, Dr. Steve Dance, one of the leading acoustics specialists in the UK, has been invaluable in giving me insight into understanding how and why my designs work in a given environment and this knowledge has further influenced my designs. If unusual problems arise, then I will work with Dr. Dance or other acoustics specialists to create an appropriate solution.

What is on the horizon?

A number of new patterns and colours are on the way as well as more designs for acoustic curtains and room dividers!

ANNETTE DOUGLAS

Annette Douglas is a third-generation textile designer based in Switzerland. In 2011, she launched the 'Silent Space Collection,' the world's first product line of transparent acoustic curtain textiles, which was awarded the Swiss Design Award in 2011, the Red Dot Best of the Best in 2012, and the Dwell on Design Award in 2012.

The Silent Space Collection grew out of a collaboration with EMPA, a research institute of material science. Can you talk about this collaborative process?

I knew EMPA would be the best partner in this field. We started with brainstorming meetings between their textile and acoustics departments. Finally in 2009 we applied for a research grant from The Swiss Commission for Technology and Innovation, and then worked for over 2 years in a small and very efficient team. We spent a lot of time in research, in materials and their acoustic properties, in construction, in testing, etc. Our goal was to create a transparent or translucent textile with acoustic absorption, flame retardancy, and last but not least a nicely designed product.

You also worked with Weisbrod-Zürrer as an industry partner who helped produce specially woven samples for acoustical testing. Was this an iterative process?

I had worked with the company for more than 10 years as a freelancer so the processes there were very familiar to me. The sampling happened in close collaboration with the technicians and Oliver Weisbrod.

Is the fabric woven in a manner to achieve a particular air resistivity to provide sound absorption?

It is a mix of various components. In our research our focus was to discover the significant components in order to define a recipe. There are other components like the porosity, the weight, the finishing, etc. — all factors contribute to good sound absorption.

'Carmen' translucent acoustical curtains are used in front of glazing throughout the atrium.

All of the textiles in the Silent Space Collection use Trevira CS Polyester. Was your decision to work with this material based on aesthetics or functionality?

Yes, we focused on Trevira CS Polyester because of functionality. It is a well-refined product particularly if you want to reach the highly demanding flame retardant properties and sustainability standards for the contract market.

Was translucency a key element of the design intent?

Yes, translucency was from the beginning my intent of the product. Working with textiles and acoustics I noticed that there were only heavy and dense fabrics on the market which block light. So a light-filtering, daylight curtain or drape with good absorption properties was totally absent on the market.

As a textile designer I would assume that aesthetics are at the forefront of your practice. Did you feel aesthetically limited by the acoustical performance requirements in developing these materials? Did you ever feel there was a trade off between the look of the material versus the sound?

Personally I prefer to work on projects which feel, as a starting point, limited. It is an interesting and challenging process to design a good product out of limitations. In the research process I always had the design, the look, the touch, the appearance, the colours, the user in my mind. We dropped ideas which may have been interesting from an acoustic perspective but not from the design or user perspective.

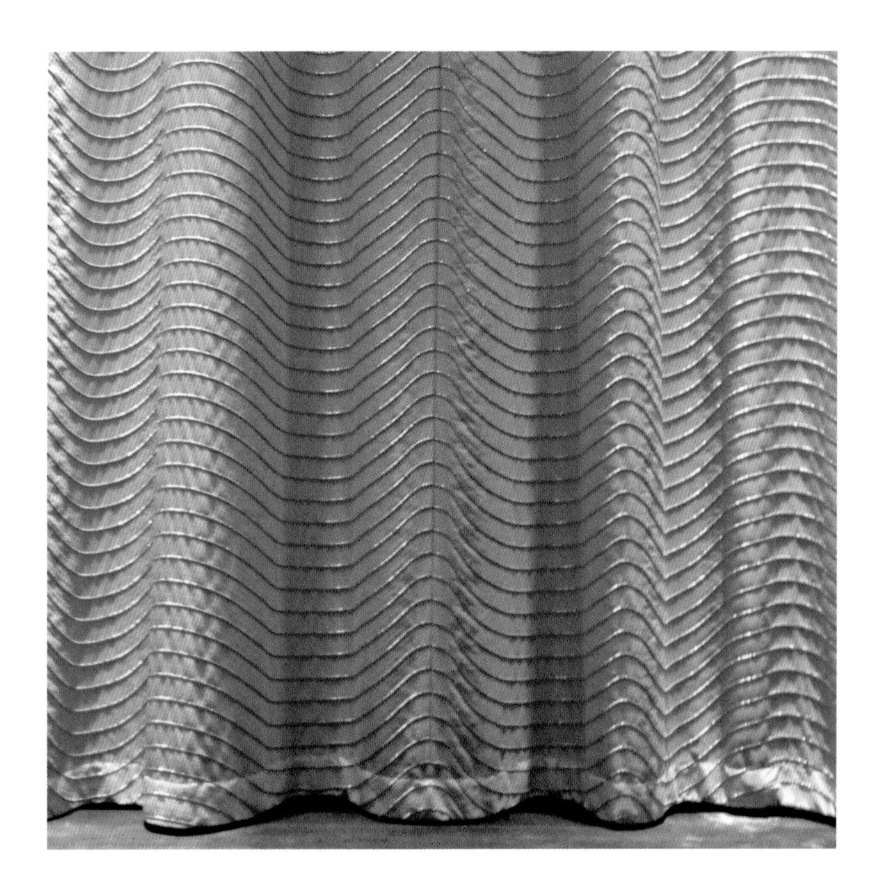

Custom embroidered acoustical curtain designed for the Toni Areal concert hall, Zurich, Switzerland.

In 2014 you expanded the Silent Space Collection. Will you continue to develop materials for this line?

We worked on the extension of the Silent Space Collection in collaboration with EMPA. We realised that we can improve the absorption properties. The line STREAMER with STREAMER classic and STREAMER pro now reach up to 80 per cent absorption; WHISPER air is our lightest and most translucent fabric which absorbs up to 60 per cent.

Our strength and main focus is on acoustic projects, such as developing new acoustic products for our own brand or for other brands, or tailor made acoustic projects for buildings which we develop in close collaboration with acoustic engineers and architects. For example, recently I received an ambitious request for a curtain for a concert hall at a university campus in Zurich. The acoustic engineers calculated the performance requirements, in terms of the absorption at various frequencies, and I had to develop the curtain to meet these requirements. Additionally, there were tight restrictions concerning flame resistance and of course many more limitations from the architectural side. The basis for this development was a fabric out of my collection 'Silent Space' and it is embroidered. The design works as tromp l'oeil. These types of projects fascinate me — when you are able to design a nice product and meet all of the targeted functional requirements. The first feedback from the acoustic engineer I received was that the testing has shown that the performance of the curtain is as they calculated in their model. And we got all certificates from the testing institute for the fire requirements.

What are you currently working on?

We just finalized a project for the Musikinsel in Rheinau, Switzerland — a hotel and retreat for musicians. We worked with the acoustic engineer Ekhard Kahle from Brussels and in close collaboration with the architects Bembé Dellinger and the artist Beat Zoderer. We were in charge of all textiles in the music hall — a combination of design, acoustics, and art. Beat Zoderer did the artwork and he wanted, if possible, to have the acoustical panels embroidered. I had to translate the artwork into an industrial embroidery process. Along the window front we put STREAMER classic out of my Silent Space Collection. The feedback from the musicians is great — they are impressed by the performance of the music hall and especially the translucent curtain which keeps the daylight in for their practice sessions. The hall can accommodate up to 80 people for orchestral rehearsal. ⁘

A glass walled meeting room with 'Liquid' model translucent curtains from the Silent Space Collection.

ERWAN BOUROULLEC

Ronan and Erwan Bouroullec are brothers and designers based in Paris. They have been working together for about 15 years, bonded by diligence, and challenged by their distinct personalities. Their work has covered many fields ranging from the design of small objects such as jewellery to spatial arrangements and architecture. The designers also pursue an experimental practice through Galerie kreo, which is essential to the development of their work.

Your first project dealing with acoustics was the North Tiles (2006) designed for Kvadrat. Were acoustics your design intent or simply a byproduct?

Kvadrat wanted us to design a showroom and since Kvadrat is a textile company, we started to think about making walls out of textiles or creating something that was in-between a curtain and a structural wall. Step by step we approached this tile concept which is a kind of sandwich of textile, foam, and textile again.

Opposite: A showroom for Kvadrat in Stockholm, Sweden, featuring The North Tiles (2006), composed of thermo-compressed foam and fabric.

Left: Kvadrat Clouds (2009) modular system of thermo-compressed foam and fabric, connected with rubber bands.

So we built the showroom walls out of these tiles, which were also hanging from the ceiling and interconnected to one another. We knew that it would probably have a positive effect on sound, but it was not the initial intent and we never studied with engineers or anything like that. The day of the opening, there was another showroom next door which had the same spatial volume and as soon as you entered the Kvadrat showroom, it was like you had changed all of the parameters on the amplifier and everything was softer and better.

In 2009 you launched Clouds. Was this an extension of the North Tiles project?

Well, it was kind of simple — the North Tile was way too expensive in the end. People are using Clouds a little bit like putting in some plants — it's more of free element that can be quite small, it doesn't need to be wide and spread out.

You introduced the Alcove sofa the same year as the North Tiles. These works seem to initiate a theme in your work of carving

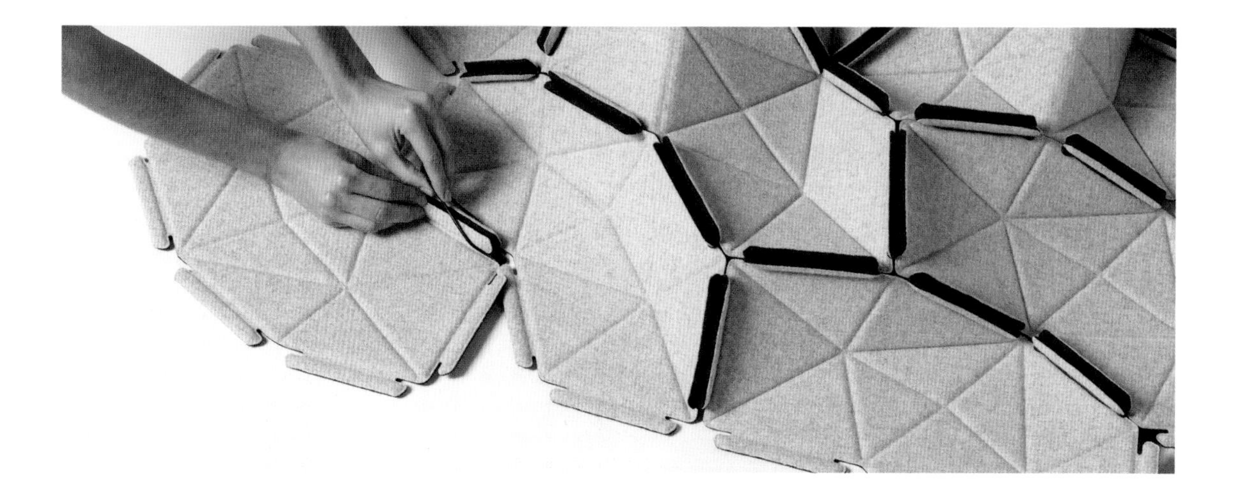

Above: Detail of Kvadrat Cloud assembly.

Opposite Above: Alcove Highback Sofa (2007).

Opposite Below: Workbays (2014).

intimate spaces from a larger public space. It's not simply visual though, the work also creates acoustic intimacy.

Regarding the Alcove sofa made for Vitra, it's the very first sofa with raised walls for contemporary times. Of course this has been in history, you can always find a sofa like this, but this one is really the first recent one that is made for intimacy, for people to sit inside and disappear within an open space.

I think the real opportunity with acoustics is the ability to change the manner or comportment of people. Behind the shape, behind the language of the material, behind everything, we try to provoke a certain gentleman attitude which has a lot to do with sound in general. Design is super important to make you behave properly. The more you can in a way temper your comportment, the better it will be for the overall sound environment.

We met some Swiss police who told us they had been buying Alcove sofas to debrief victims that were under stress because they found that this kind of sofa, the kind of environment it creates, is much quieter for them and helps the victims open up. I think that our design practice tries to create environments that give solutions to people with which there is a certain dignity.

Your Workbay series is another example of spatial subdivision. Is this design a response to the current trends in open-plan office environments, which seem to privilege transparency, density, and openness in favor of privacy?

Well, it's really a complement to open space, which is a dynamic space, but inside the open space you need enclosure. A lot of companies work in open spaces because they may not have the resources to create all of these rooms. It's much quicker to have an open space and then all of the spatial organiation can be made with furnishings. We are product designers so most of the time what we create is based on the idea of the object of furniture. So while furniture doesn't really belong to the space, it is mobile and can be positioned as you like. ⁘

FORM US WITH LOVE

Form Us With Love is an international design studio headquartered in Stockholm, Sweden. Their work falls within three categories: Consultancy—an engagement in products, ranges, collaborations and spaces for clients around the world, Ventures—disciplined and holistic approaches to launch and build brands, Civic—based on knowledge transfers and sharing of experiences, actively contributing to the broader spectra of design. In 2014, Form Us with Love co-founded BAUX, seizing upon an opportunity to reinvent the aesthetics of a function-heavy Swedish-made building material, Träullit.

How did your development of Baux come about?

One day, we were having lunch at an art museum not far from our new studio. On the wall of the restaurant, the interior architect had placed sound absorbing panels. But not just any panel, a specific type of grey square that we knew from our childhood that covered the ceilings in schools, gyms, etc.

Most people who grew up in Sweden know this panel. It might not be something that people necessarily pay attention to, but the fact that it is used by all governmental institutions and most office contractors, makes the chance of a Swede ever seeing

BAUX rectangular panels with diagonal patterned surface.

one, very high. The material was rough and smooth at the same time, something we thought would be suitable for the general feel we had created in our studio. We sent for samples and received two types of grey. Immediately we began experimenting — 'this is something we can add a little colour to,' we thought.

Our passion for the material grew and we decided to visit the only factory in Sweden which produced the panels — the family-owned company called Träullit. Upon our arrival we met the owner Bengt Rääf. He showed us around and spoke passionately about the material they produced, wood wool, a mix between cement, water and wood. A year later we launched the brand we call BAUX.

Wood wool products such as Träullit have been around for many decades. What are some of the advantages and challenges of this material?

Wood wool is sound absorbing, fire resistant, moisture regulating, heat distributing, and made using natural materials. One of the benefits is that it is a well-known material that has been around, so we know from the start that the features and characteristics of the material were really good. One challenge was to make architects understand that

Above: BAUX 3D Pixel Tiles

Opposite Above: BAUX Hexagon Tiles

Opposite Below: Hood (2012) designed by Form Us With Love for Ateljé Lyktan — a modular and easy to assemble pendant lamp made from moulded polyester.

we were creating a framework for them and not as it had been before, something that a builder installed in the last minute to reduce sound pollution. So instead of having a standard sheet in two shades of grey, we created a wide range of geometric tiles in a number of sizes, and as a start we offered 25 vibrant colours, and all of a sudden the possibilities are endless.

How did you arrive at the core dimensions and shapes for your tiles?

It was important to get an intuitive feel when you see the range of tiles, that you understand it's a system. The sizes range from 30 cm to panels that are 120 cm. The idea is that large scale is more suitable for bigger spaces and vice versa.

Following the panels and tiles, BAUX launched the PIXEL line. What were the motivations for this product line?

For this, our third collection, we spent time investigating the thickness of the panels being produced. Previously we had only been looking at one standard thickness and, studying the production, we realised that altering the thickness was an easy move to apply. In addition, we found that building a sound absorbing wall with alternating heights and thicknesses, created a better sound quality and a greater sense of dimension for the overall interior experience.

What is next on the horizon for BAUX?

Form Us With Love is in the business of doing things better, so we always have our minds set on developing things. We have been intrigued by making office spaces better and smarter since we started our business. When it comes to sound absorbing solutions, we have been busy with solutions such as Hood for Ateljé Lyktan and Plaid for Abstracta. ⁘

CHRIS DOWNEY

Christopher Downey is an architect, planner, and consultant who lost all sight in 2008. Today, he is dedicated to creating more helpful and enriching environments for the blind and visually impaired. Chris consults on design for the blind and visually impaired, encompassing specialized centers as well as facilities serving the broader public. His work ranges from a new Department of Veterans Affairs blind rehabilitation center, to renovations of housing for the blind in New York City, and to the new Transbay Transit Center in San Francisco. He also teaches accessibility and universal design at UC Berkeley and serves on the Board of Directors for the Lighthouse for the Blind in San Francisco.

We often consider architectural design practice as visually driven. With the loss of sight, how has your approach to architecture changed?

On a sort of grand scale, it's taking the idea of architectural design and aesthetics from the visual to the multi-sensory. It's really thinking through architecture to be experienced in more ways than just how you see it. Typically when architects design things, they're really trying to design for the eye and there are a lot of things that you can do to share that level of aesthetic and design for more than just the eyes.

You recently consulted on the design for the new headquarters of San Francisco's LightHouse for the Blind and Visually Impaired, which opened in June 2016. What can you tell us about this project?

The LightHouse for the Blind reception lobby. The concrete floor guides visitors through the building. A wood slat acoustical ceiling helps reduce noise at the reception desk and waiting area.

The LightHouse for the Blind is the oldest and largest agency in Northern California providing training, information, and advocacy for blind and visually impaired people - and one of the largest in the country, serving 3000 people a year. The new facility is 3 storeys of 3700 m^2 of space in a high rise building in downtown San Francisco, which supports their classes and programs and includes dorms for students enrolled in the centre's training program.

I worked with Mark Cavagnero Associates, consulting on everything from lighting strategies and levels, colours, materials, acoustic design, and tactile elements that we know people are going to grab like handrails and doorknobs. One of the main design strategies was to find a way to express a sense of the overall space — that there is more than just the one floor when you first enter. By having an open stairwell, you can visually see that there is this spatial organization and connection between floors, and by designing the stair treads out of wood, you can now hear footfalls and canes going up and down the stairs and this organisation can be perceived by someone who is blind. If you only had the elevators to rely upon, you would have no idea that more than one floor existed unless you had been there before.

Each floor is a long, rectilinear floor plate and the stairwell helps break that long rectangle into two ends. So whenever you're moving around on any of those three floors, as you return to the middle there is the open stairwell, which serves as an acoustic landmark. We wanted

Right: The open stairwell serves as an acoustic landmark and allows visitors to hear the connectivity between floors.

Opposite: A circulation loop with a concrete floor and drywall ceiling helps cane taps resonate and establishes an auditory blueprint for each floor.

the space to be open, welcoming, exciting, and dynamic. Especially for those that are new to sight loss and visiting the building for the first time, it can be a pretty anxious experience.

What are some of the acoustic design aspects of this project?

Acoustic design, especially for office spaces, is typically dealt with in terms of isolating sound from one space to another or controlling mechanical noise. Our objective was to design the quality of acoustic experience. Some of it is on a very simple level like the circulation diagram, which for each floor is a rectangular loop around a small central lobby where the elevator and stairwell are located. The circulation loop has a polished concrete floor and it's the only place you find this. As people walk, you can hear the footfalls, tapping or the sliding of canes and that animates the space so you can hear others even if you can't see them down the hallway. You might even recognize the sound of someone's cane — everyone uses their canes in different ways and they sound differently — so you can hear it's your friend down the hallway and you could call out to him. It also has the benefit of not sounding like a ghost town. You can't see anybody, so it's nice to know there are people and activities around you. So that circulation loop is designed to be acoustically live but not so much that it becomes disturbing.

One of the ways we accomplished this is working with an acoustic design consultant that can model the sound of a space when it's in the design phases using the digital building information models (BIM) and assigning the acoustic attributes for material, form, and volume. Designers can use ray tracing to show how light bounces around space - you can do the same thing with sound and hear the space as it's being designed. So we spent a lot of time with the client and ourselves working on adjusting the acoustic quality.

Having a concrete floor in a circulation space is sometimes counter to certain ideas about noise control. For instance, carpet might normally be used to quiet footfall noise.

I'm working on another project about airports, and I'm advocating for the use of hard surfaces throughout the main concourse area of the airport terminals. There are airports that just run carpet from wall-to-wall everywhere you go and you can't hear a thing. You don't know where on earth the space goes. The ceiling heights are probably all the same and with carpet you lose everything. It just sounds like a flat dead environment with no acoustic clarity or structure to the space. You have no idea where to find the gates or where the concourse is even flowing. By having the hard floor surface all of a sudden you can hear footfalls further ahead so you can target that area as you walk.

For the LightHouse, there is carpet used in many adjoining spaces. To assist in navigation, we designed the transitions from the concrete floor circulation loop with metal strips. Where you have two different flooring materials, there's usually a transition strip and rather than have that floor transition be made of rubber, which doesn't give you a lot of information when you hit it with a cane, we used a metal transition strip so that when the metal tip of the cane hits it, it makes a good crisp sound. It's an easy edge to work with that can be felt through the end of your cane and can be clearly heard.

You mentioned designing to a level of aesthetics for more than just the eyes. Can you provide another example from the LightHouse project?

On a purely aesthetic level was the handrail design. If you're blind, you don't go around exploring things randomly with your hands – unless you're a curious architect like me. So I don't think about designing most wall surfaces as tactile. But there are those places that we know you're going to reach out and interact with physically in the building. One of those most critical places was the stairwell and the handrail. I really wanted to design the handrail to fit the hand in a very unique and different way, not for the sake of being different, but in a very generous way that struck you.

For this project, I initially worked on the design using wax sticks. It's a tactile way for me to draw. I created different profiles with the wax sticks and then Mark Cavagnero Architects created hand drawings and then 3D prints of the various handrail options that we could explore through touch and share with the client. It's all about the surprising delightful manner in which your hand can grip the handrail. It's showing a level of care in the design in ways that are very intentional and interactive rather than, for example, just putting in a simple pipe rail. ⁘

BRADY PETERS

Brady Peters is an architect and designer based in Toronto, Canada, and Assistant Professor at the John H. Daniels School of Architecture, Landscape, and Design. He specialises in the architectural applications of computer programming, the integration of digital fabrication in the design process, and the simulation and design of performance-driven forms. Brady was an Associate Partner at Foster + Partners with the Specialist Modelling Group, the office's internal research and development consultancy. He has previously worked in the London office of design-led engineering practice Buro Happold. Brady is also the director of the Smartgeometry Group.

Your work and research explores parametric design tools, computation, and digital fabrication in architectural contexts. How do these technologies intersect with acoustics?

I left Foster + Partners in 2008 to undertake my PhD research at the Centre for Information Technology and Architecture (CITA) which is a research unit within the Royal Danish Academy of Fine Art's School of Architecture. In my research I collaborated with local architects and engineers on the problem of how to bring sound into the building design process. Architectural design is now pretty much exclusively carried out in the digital realm, and the digital offers some opportunities to designers that we didn't have when we were just working with pencils and drafting boards. Computers can simulate

Opposite: Distortion II (2011) is an experimental research project, designed, built, and tested to create visual and acoustic affects within an open-plan space. Through its geometry and material, a complex wall surface creates spaces with different acoustic characteristics. The project was designed specifically to probe two acoustic extremes: a sound-amplified zone, and a sound-dampened zone. Custom software was developed to allow for real-time analysis of acoustic parameters while the structure was modelled. Absorptive and reflective material properties were mapped to achieve the desired performance and then output to a CNC or laser cutter for final production. Distortion II was designed by Brady Peters, Martin Tamke, Anders Holden Deleuran, and Dave Stasiuk.

Right: Detail of Distortion II surface showing absorptive and reflective materials.

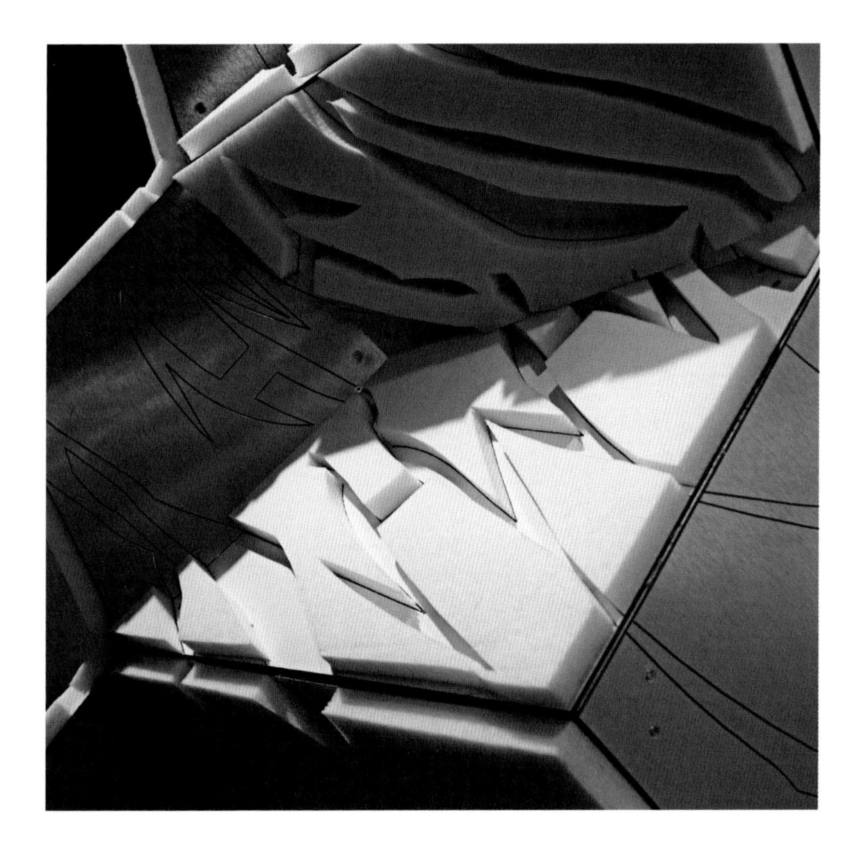

aspects of performance that were very previously challenging for architects to consider – aspects such as daylight, air flow, thermal, solar, energy, and, of course, sound. As a designer with a background in music, and skills in simulation and the generation of complex 3D forms, I was very inspired to find myself working in a research area at the intersection of architecture, sound, and the digital.

In my early work at Foster + Partners I was writing a lot of computer code using the newly available scripting environments in architecture computer-aided design (CAD) software. At first I used this capability to generate a whole lot of complex geometry – large space frames and skyscrapers; but I found that this technology enabled both better communication between disciplines but also the integration of performance into the design. For example, I could generate data in many different formats so I could produce the models required for structural analysis or the data sheets necessary for fabrication planning. Simulation could also be used to drive architectural designs; for example, solar analyses could inform the size of shading and aperture elements in a facade.

One of your earliest projects was the undulating glass ceiling that encloses the Kogod Courtyard in Washington DC, designed by Foster + Partners. This ceiling is not only geometrically complex but has also been cleverly designed with integrated acoustic absorption. Was the acoustic absorption an integral part of the ceiling's design?

Of course there were many different performance considerations that factored into the design — structure, light, and sound were all critically important factors. As the courtyard space was to be used for gatherings and musical performances, the reverberation of sound needed to be controlled. Working with acoustical consultants Sandy Brown Associates, an optimal reverberation time of about 3-4 seconds was determined. As the building was a historic landmark, making changes to the walls of the courtyard space was not allowed, and as the courtyard floor was stone and the ceiling was glass this left little opportunity for acoustic absorbing material! Our solution here was to combine structure and sound. The structural beams became acoustic absorbers - they were faced with sound absorbing material. We designed a custom material of recycled textile faced with thin aluminium rods and this was tested in a laboratory. The measured results were fed back into the parameters of our computer model. I wrote a single computer program that generated the entire structure. In the algorithm, the depth of the beams was related to the required reverberation time, the size of the space, and the performance of the acoustic material.

Some of your recent research activities, such as the Distortion Projects, have involved creating more nuanced acoustic environments. This work seems to explore iterative processes that utilise acoustic simulation software, parametric modelling, and digital fabrication. What do you envision as the ultimate objective of this exploration?

I believe that how a space sounds is as much part of its architectural experience as how it looks. I don't mean that different sensorial experiences are necessarily equal, but I think it is important to recognise that different senses - in particular sight and sound - contribute to the overall impression of an architectural experience. My research promotes a view that sound is more than simply utility - it is not just about efficient communication of speech or the removal of environmental noise - but that designing for sound has a creative potential that architects have yet to really explore. In my PhD, a lot of my background research showed that most architects are genuinely interested in sound and its design potential, but they don't have the tools to explore this territory. To address this, my research attempted to create some new design tools and workflows for building designers to engage with sound in a creative way. The two Distortion experiments in Copenhagen and the FabPod experiment in Melbourne attempted to create acoustic experiences that varied as one navigated through the space of the installation. Through the manipulation of surface geometry and material the acoustic properties of a space were modified, and the consequences of these manipulations were simulated during the design phase. In each project, I developed bespoke computational tools and design workflows that brought together designed aural experience, surface geometry, and material performance. A key part of the evaluation of these experiments was their realisation at 1:1 scale. To do this required the use of a lot of different types of digital fabrication machinery. The production of the 1:1 prototype also means that its acoustic qualities could be experienced. ⁘

Opposite: FabPod (2013) is an acoustically tuned meeting room housed within an open-plan office at the RMIT Design Hub The form and materiality arose from a Smartgeometry workshop in 2011 that investigated the acoustic properties of hyperboloid geometry. By positioning hyperboloids to maximise sound scattering within the space, the FabPod has an unusually warm and diffuse acoustic. The FabPod Project was designed by Mark Burry, Jane Burry, Nicholas Williams, Brady Peters, Daniel Davis, and Alex Pena de Leon.

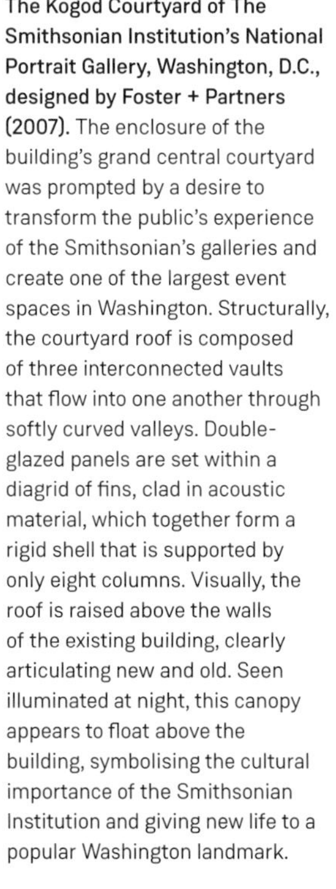

The Kogod Courtyard of The Smithsonian Institution's National Portrait Gallery, Washington, D.C., designed by Foster + Partners (2007). The enclosure of the building's grand central courtyard was prompted by a desire to transform the public's experience of the Smithsonian's galleries and create one of the largest event spaces in Washington. Structurally, the courtyard roof is composed of three interconnected vaults that flow into one another through softly curved valleys. Double-glazed panels are set within a diagrid of fins, clad in acoustic material, which together form a rigid shell that is supported by only eight columns. Visually, the roof is raised above the walls of the existing building, clearly articulating new and old. Seen illuminated at night, this canopy appears to float above the building, symbolising the cultural importance of the Smithsonian Institution and giving new life to a popular Washington landmark.

PHILIP ROBINSON
FOSTER + PARTNERS

Philip Robinson is an Environmental Design Analyst at Foster + Partners, an international architecture and integrated design studio. He is a member of the Specialist Modelling Group, which combines two principal areas of expertise: environmental analysis and simulation; and parametric modelling and fabrication of complex geometries. His role encompasses analysis and simulation of acoustic issues at every scale, throughout the design process. Prior to joining the Specialist Modelling Group, he conducted research on the perception of concert hall acoustics. He holds a PhD in Architectural Science from Rensselaer Polytechnic Institute.

It is rare for architectural firms to staff their own acoustical engineers. What are the benefits of having an in-house acoustician?

The acoustics work we do is not intended to replace consulting from external acoustic firms; in fact, most of our projects do have an external consultant. There are two areas where we expand on the work of the external consultants.

The first is to find integrated solutions to the technical requirements that consultants provide. This may involve working with the other disciplines in our group to find synergies. We have specialists in thermal comfort, lighting, airflow, structures, and complex geometry, which allows us to examine design problems in a holistic way; this interdisciplinary mode of working is much more difficult when the architect relies on individual specialised external consultants. Having this expertise within the architectural practice makes conversation with the consultants much more two-way. Rather than grudgingly accepting what the consultant recommends, we can collaborate to come to an integrated solution. For example, we were recently asked to conceal loudspeaker arrays behind a stretched fabric ceiling soffit. The soffit material had to be acoustically transparent for acoustic purposes, but also had to be airtight to control airflow, an appropriate opacity to control natural light, and have some thermal resistance for passive cooling requirements. Having the ability to measure materials in-house and run multi-disciplinary simulations of all the environmental conditions ourselves allowed us to quickly come to a solution.

The second area is more about helping clients and architects to understand architectural acoustics. We have found that numbers and graphs are pretty ineffective at communicating the acoustic quality of a space. Even if someone has a really good understanding of what decibel levels or reverberation times mean, which is rare, these metrics still don't tell you much about the quality of a space. For this reason, we have started doing a lot of auralisations, presentations using head-mounted virtual reality systems, and augmented reality apps on tablets and phones. We recently completed construction of a 12 channel auralisation suite in our office, so we can now do fully immersive sound field simulations. This is great not only for presenting to clients, but also helping the architects and designers understand the effect of the materials they are selecting. This allows us to explore design options much more freely than if we relied on consultants for auralisations. Every office has people that produce visualisations, it is only natural that auralisations should have a place in the process as well.

In terms of materials, one would associate Foster + Partners projects with glass, metal, and concrete. How are acoustical considerations integrated without compromising the aesthetic vision of the project?

It is true that the firm's material palette tends towards pretty reverberant spaces, and that the aesthetic vision can be at odds with acoustic quality requirements. On one front, we are attempting to raise the importance of the acoustic quality. The auralisation suite has a big role in this because when the designers hear how loud and reverberant it is, and realise that they can't speak to each other at close distances, they become more committed to incorporating acoustic treatments. On the other hand, large areas of applied treatment are never acceptable. The acoustic treatments must be integrated; in the end they generally don't look like acoustic treatments. Usually the material doing the work is mineral wool, but we work very hard to conceal it into the finishes. We do a lot of simulations and testing to determine if a particular concealment arrangement will provide adequate performance, and adjust accordingly. We've done this for metal light diffusors at the ceiling, new arrangements of wood lattices, and for some very fine metal screens.

What are some current material innovations being developed by the Foster + Partners acoustics division?

Mineral fibre is doing most of the absorption. The innovation is how to incorporate it in the best way. One way is with acoustically transparent materials, which are also useful for concealing loudspeakers. There is little data available on this and we are building a database of material transparency measurements so we have a larger selection of materials to choose from. ⁘

JOHN COAKLEY
CARBON AIR

John Coakley is CEO of Carbon Air, a small company with a big list of technologies in development in acoustics and beyond, all based on the effects of material geometries that are invisible to the naked eye.

Carbon Air has developed an emerging material technology using activated carbon, which absorbs sound in a rather unusual way. What is activated carbon and how is it made?

Activated carbon is made by baking organic materials (wood, coconut shell, peat, coal, and others) until all matter is converted to carbon. It is then 'activated', usually by driving water vapour through the material at high temperature and pressure to excavate the former cell cavities, revealing a delicate and highly porous carbon skeleton featuring millions of microscopic pores.

The functional origin of these pores means they fall into an organised hierarchy with broad (mesoscopic)

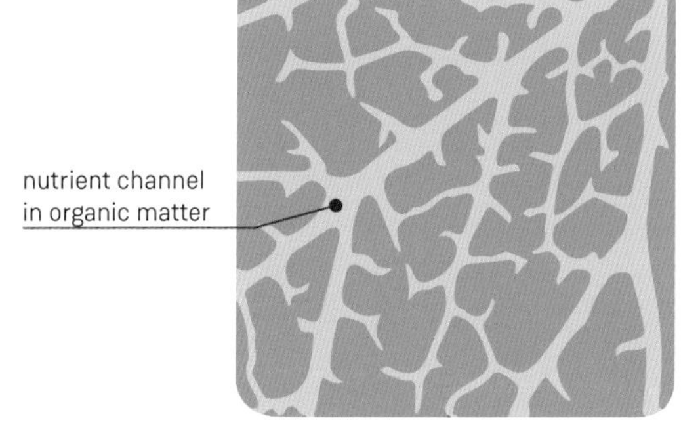

ORGANIC MATERIAL

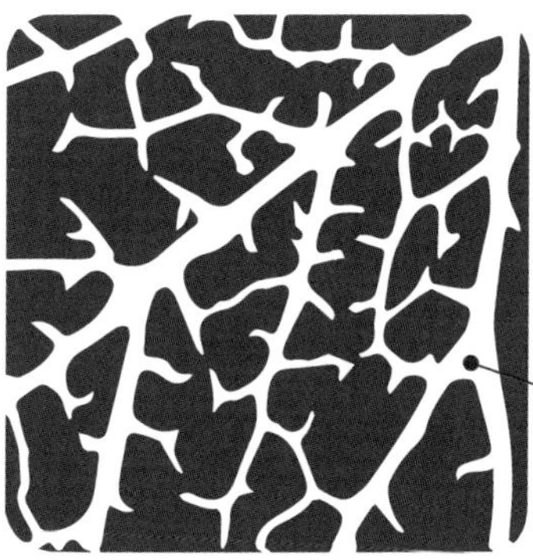

ACTIVATED CARBON

transport channels feeding smaller (microscopic) chambers which themselves feed into millions of miniscule (nanoscopic) ones. It's a bit like a fractal.

The material comes in a range of different forms — from fine powders to large granules — and a range of different activation levels. But it's always made up of just one thing — pure carbon.

How does activated carbon differ from a traditional porous material?

There are a couple of things about activated carbon that are special. The first is the very high surface area, amounting to over a thousand square metres per gramme, in the case of the good stuff. The material's porosity also occurs at a wide range of discrete scales, which is important for noise absorption.

But the second and more miraculous aspect is the thing that happens at the smallest scale; a thing called 'adsorption/desorption'. This allows the material to effectively 'breathe in' implausible amounts of air when pressure rises or temperature falls. In fact, a cup full of activated carbon will contain about six times as many air molecules as a cup containing air alone. It's a very pronounced effect.

How does that work?

All materials support adsorption/desorption to some extent, but activated carbon does it very powerfully.

Air molecules are actually repelled by the electronic forces present in the face of a material. But they are also attracted by another force — the 'Van der Waals' forces — that seek to bind the molecules to the surface. These are stronger than the electronic forces over a larger distance, resulting in a zone being created above the surface of a material that effectively captures air or gas molecules.

The nanopores of activated carbon multiply this effect several times, because the geometry of the pore cells coincides nicely with the height of this Van der Waals attractive zone — but with each face on each side of the cell exerting its own pull. So the heart of the nanopore cell becomes a hotspot of densely captured gas molecules.

When the gas molecules are captured this way, they pack much more tightly than they would in the open, but without adding to pressure. In an unscientific sense, they disappear, at least in terms of their contribution to pressure.

How does this technology work as a sound absorber?

At Carbon Air, we believe we are the first to fully understand the acoustic consequences of adsorption/desorption in multi-scale porous materials, and to capture this behaviour in a numerical model that we use as an optimising design tool.

There are a number of factors at work, some involving the range of different scales of pore involved, and the pressures that are experienced at each, and the presence of adsorption at the nanoscale sites. But the net effect is that these materials absorb much more sound energy than normal materials across the range, but particularly at low frequencies.

Typically a conventional porous absorber needs to be very thick to attenuate low frequencies because wavelengths are long and as a rule of thumb, the absorber needs to be about a quarter of the wavelength deep to be effective. But an activated carbon layer can achieve the same attenuation in much thinner layers, in part because it changes the mechanical or acoustic compliance of the air trapped inside the layer — meaning that the air behaves as if it has greater depth.

The activated carbon is usually integrated within another material. Why is this?

Activated carbon powder is a bit like the powdered paint you played with at primary school — it gets everywhere. And even the better granular form gives off some powder over time. It's also quite expensive for a building material. So our challenge has been to find ways to integrate it into other materials without masking the pores and blocking their acoustic action.

We've also been exploring other materials that have some of the features of activated carbon but are much cheaper, such as perlite.

What are some of the acoustical applications on the horizon for Carbon Air?

We're working on an array of slender acoustic elements that can sit beneath concrete ceilings that are being used as thermal mass in a building scheme. We've shown in the laboratory that with the right spacing, the array becomes thermally 'invisible', allowing the transport of heat energy into and out of the thermal mass, while still absorbing noise using the array's thin layers. This, we believe, will be a really useful breakthrough in applications like green office spaces and school buildings where use of thermal mass is key to reducing CO_2 emissions by naturally reducing diurnal temperature swings, but where acoustic intelligibility is also extremely important. Normal porous absorbers are acoustically insulating and so will block the required thermal transfer.

Aside from building acoustics, we are developing industrial partnerships for acoustic trims in high-end automotive and truck applications. We have worked on air intake silencers and HVAC linings that can benefit from the same behaviour, and we see big opportunities as our optimised configurations become thinner and more robust.

We are also exploring applications in audio enhancement — something that has been known about activated carbon for years. We have shown how the material improves the bass response in sub woofers, making a small box behave like a bigger box, and we have captured this in numerical modelling.

One goal is to create a material that can be applied into architectural, automotive, and industrial noise proofing that can seamlessly double up as an audio enhancement lining. We believe we're already very close. ⁘

APPENDIX: ABSORPTION COEFFICIENTS

The tables provide absorption coefficients in octave bands for everyday building materials and common porous materials.

	SOUND ABSORPTION COEFFICIENTS OCTAVE BAND FREQUENCY [HZ]						
MATERIAL	125	250	500	1000	2000	4000	**NRC**
Brick, exposed, unglazed, unpainted	0.03	0.03	0.03	0.04	0.05	0.07	**0.05**
Concrete block, coarse, unpainted	0.36	0.44	0.31	0.29	0.39	0.25	**0.35**
Concrete block, painted	0.10	0.05	0.06	0.07	0.09	0.08	**0.05**
Concrete, poured, unpainted	0.01	0.01	0.02	0.02	0.02	0.03	**0.00**
Linoleum or vinyl stuck to concrete	0.02	0.02	0.03	0.04	0.04	0.05	**0.05**
Wood floor boards on joist floor	0.15	0.20	0.10	0.10	0.10	0.10	**0.15**
Wood platform with large airspace below	0.40	0.30	0.20	0.17	0.15	0.10	**0.20**
Parquet fixed with asphault on concrete	0.04	0.04	0.07	0.06	0.06	0.07	**0.05**
Solid Wooden Door	0.14	0.10	0.06	0.08	0.10	0.10	**0.10**
Glass, ordinary window glass	0.35	0.25	0.18	0.12	0.07	0.04	**0.15**
Single pane of glass, 3 mm thick	0.08	0.04	0.03	0.03	0.02	0.02	**0.05**
Glass, large panes, heavy glass	0.18	0.06	0.04	0.03	0.02	0.02	**0.05**
Double-glazing, 2-3 mm glass separated by 1 cm air gap	0.10	0.07	0.05	0.03	0.02	0.02	**0.05**
Gypsum Board, 12 mm on studs	0.29	0.10	0.05	0.04	0.07	0.09	**0.05**
Plaster, 22 mm, gypsum or lime on brick	0.01	0.02	0.02	0.03	0.04	0.05	**0.05**
Plaster, 22 mm, on concrete block	0.12	0.09	0.07	0.05	0.05	0.04	**0.05**
Plaster, 22 mm, lath on studs	0.30	0.15	0.10	0.05	0.04	0.05	0.10
Wood, 25 mm thick, paneling with air space behind	0.42	0.21	0.10	0.08	0.06	0.06	**0.10**
Carpet, heavy, no underlayment, on concrete	0.02	0.06	0.14	0.37	0.60	0.65	**0.30**
Carpet, heavy, on foam rubber underlayment	0.08	0.24	0.57	0.69	0.71	0.73	**0.55**
Carpet, heavy, with impermeable latex backing on foam rubber	0.08	0.27	0.39	0.34	0.48	0.63	**0.35**
Carpet, thin, indoor-outdoor, no underlayment	0.01	0.05	0.10	0.20	0.45	0.65	**0.20**
Audience in upholstered seats	0.39	0.57	0.80	0.94	0.92	0.87	**0.80**
Unoccupied well-upholstered seats	0.19	0.37	0.56	0.67	0.61	0.59	**0.55**
Chair, metal or wood seat, unoccupied	0.15	0.19	0.22	0.39	0.38	0.30	**0.30**
Light velour curtain, 338 g/m², hung flat on wall	0.03	0.04	0.11	0.17	0.24	0.35	**0.15**
Light velour curtain, 338 g/m², hung in folds on wall	0.05	0.15	0.35	0.40	0.50	0.50	**0.35**
Medium velour curtain, 475 g/m², draped to half area	0.07	0.31	0.49	0.75	0.70	0.60	**0.55**
Heavy velour curtain, 610 g/m², draped to half area	0.14	0.35	0.55	0.72	0.70	0.65	**0.60**
Ballast or crushed stone, 0.64 cm, 15.2 cm deep	0.22	0.64	0.70	0.79	0.88	0.72	**0.75**
Ballast or crushed stone, 3.18 cm, 15.2 cm deep	0.19	0.23	0.43	0.37	0.58	0.62	**0.40**
Ballast or crushed stone, 3.18 cm, 30.5 cm deep	0.27	0.58	0.48	0.54	0.73	0.63	**0.60**
Ballast or crushed stone, 3.18 cm, 45.7 cm deep	0.41	0.53	0.64	0.84	0.91	0.63	**0.75**
Water Surface of a Swimming Pool	0.01	0.01	0.01	0.01	0.02	0.02	**0.00**
Snow, freshly fallen, 100 mm thick	0.45	0.75	0.90	0.95	0.95	0.95	**0.90**
Grass, marion bluegrass, 50 mm high	0.11	0.26	0.60	0.69	0.92	0.99	**0.60**

DESCRIPTION	SOUND ABSORPTION COEFFICIENTS OCTAVE BAND FREQUENCY [HZ]						
	125	250	500	1000	2000	4000	**NRC**
Generic Mineral Fibre Ceiling Tile, 19 mm thick [E-400]	0.72	0.84	0.70	0.79	0.76	0.81	**0.75**
Generic Mineral Fibre Ceiling Tile, 25 mm thick [E-400]	0.76	0.84	0.72	0.89	0.85	0.81	**0.85**
Generic Fibreglass Ceiling Tile, 19 mm thick [E-400]	0.74	0.89	0.67	0.89	0.95	1.07	**0.85**
Generic Fibreglass Ceiling Tile, 25 mm thick [E-400]	0.77	0.74	0.75	0.95	1.01	1.02	**0.85**
Fibreglass Board, 16 kg/m³, 25 mm thick [A]	0.17	0.33	0.64	0.83	0.90	0.92	**0.70**
Fibreglass Board, 16 kg/m³, 50 mm thick [A]	0.22	0.67	0.98	1.02	0.98	1.00	**0.90**
Fibreglass Board, 16 kg/m³, 75 mm thick [A]	0.43	1.17	1.26	1.09	1.03	1.04	**1.15**
Fibreglass Board, 16 kg/m³, 100 mm thick [A]	0.73	1.29	1.22	1.06	1.00	0.97	**1.15**
Fibreglass Board, 16 kg/m³, 25 mm thick [E-400]	0.32	0.41	0.70	0.83	0.93	1.02	**0.70**
Fibreglass Board, 16 kg/m³, 50 mm thick [E-400]	0.44	0.68	1.00	1.09	1.06	1.10	**0.95**
Fibreglass Board, 16 kg/m³, 75 mm thick [E-400]	0.77	1.08	1.16	1.08	1.05	1.18	**1.10**
Fibreglass Board, 16 kg/m³, 100 mm thick [E-400]	0.87	1.14	1.24	1.17	1.18	1.28	**1.20**
Fibreglass Board, 48 kg/m³, 25 mm thick [A]	0.11	0.28	0.68	0.90	0.93	0.96	**0.70**
Fibreglass Board, 48 kg/m³, 50 mm thick [A]	0.17	0.86	1.14	1.07	1.02	0.98	**1.00**
Fibreglass Board, 48 kg/m³, 75 mm thick [A]	0.53	1.19	1.21	1.08	1.01	1.04	**1.10**
Fibreglass Board, 48 kg/m³, 100 mm thick [A]	0.84	1.24	1.24	1.08	1.00	0.97	**1.15**
Fibreglass Board, 48 kg/m³, 25 mm thick [E-400]	0.32	0.32	0.73	0.93	1.01	1.10	**0.75**
Fibreglass Board, 48 kg/m³, 50 mm thick [E-400]	0.4	0.73	1.14	1.13	1.06	1.10	**1.00**
Fibreglass Board, 48 kg/m³, 75 mm thick [E-400]	0.66	0.93	1.13	1.10	1.11	1.14	**1.05**
Fibreglass Board, 48 kg/m³, 100 mm thick [E-400]	0.65	1.01	1.20	1.14	1.10	1.16	**1.10**
Fibreglass Board, 48 kg/m³, FRK Foil Facing, 25 mm thick [A]	0.18	0.75	0.58	0.72	0.62	0.35	**0.65**
Fibreglass Board, 48 kg/m³, FRK Foil Facing, 50 mm thick [A]	0.63	0.56	0.95	0.74	0.6	0.35	**0.75**
Fibreglass Board, 48 kg/m³, FRK Foil Facing, 25 mm thick [E-400]	0.33	0.49	0.62	0.78	0.66	0.45	**0.65**
Fibreglass Board, 48 kg/m³, FRK Foil Facing, 50 mm thick [E-400]	0.45	0.47	0.97	0.93	0.65	0.42	**0.75**
Polyurethane Foam, 6 mm thick [A]	0.09	0.12	0.25	0.32	0.71	0.97	**0.35**
Polyurethane Foam, 12 mm thick [A]	0.12	0.18	0.37	0.65	0.95	0.97	**0.55**
Polyurethane Foam, 25 mm thick [A]	0.16	0.30	0.68	1.00	1.00	0.94	**0.75**
Polyurethane Foam, 50 mm thick [A]	0.32	0.68	0.92	0.96	1.00	1.00	**0.90**
Polyurethane Foam, 25 mm thick, Mylar Film Facing [A]	0.08	0.21	0.6	0.43	0.48	0.72	**0.45**
Polyurethane Foam, Convoluted Egg Crate, 50 mm thick [A]	0.15	0.31	0.73	1.04	1.08	1.12	**0.80**
Melamine Foam, 25 mm thick [A]	0.10	0.10	0.30	0.70	0.80	0.90	**0.60**
Melamine Foam, 50 mm thick [A]	0.05	0.31	0.81	1.01	0.99	0.95	**0.80**
Melamine Foam, 50 mm thick, Factory Painted Coating [A]	0.13	0.41	1.02	1.18	1.18	1.13	**0.95**
Cotton panel, 48 kg/m³, 25 mm thick [A]	0.08	0.31	0.79	1.01	1.00	0.99	**0.80**
Cotton panel, 48 kg/m³, 50 mm thick [A]	0.35	0.94	1.32	1.22	1.06	1.03	**1.15**
Cotton panel, 96 kg/m³, 25 mm thick [A]	0.07	0.30	0.86	1.10	1.05	1.03	**0.85**

INDEX

PHOTOGRAPHIC CREDITS

Unless otherwise identified below, photographs have been taken by the author.

P. 6-7 ©Woven Image (acoustic tiles Delta, Square and Oblong made from EchoPanel) **P. 29-30** ©Anna Kyyro Quinn Studio **P. 36-37** ©RVTR **P. 50-51** images by Brad Feinknopf/OTTO **P. 53** ©Ecophon **P. 62** ©sander architects, LLC, photo by Sharon Risedorph Photography **P. 64** ©Filzfelt **P. 66-67** ©i29 I interior architects **P. 68-69** ©HplusF, photo by Michal Czerwonka **P. 70-71** ©Freund GmbH **P. 73** ©Habitat Horticulture, photo by Gary Belinsky, wall design and installation, David Brenner, Habitat Horticulture **P. 76** Heradesign **P. 78-79** ©Kirei/ Woven Image **P. 80-81** ©Instyle Contract Textiles, photos by Fiona Susanto **P. 82-83** ©KnollTextiles **P. 85** ©Snowsound **P. 86-87** ©RCKa, photos by Jakob Spriesterbach and Ivan Jones **P. 90-91** ©Barber & Osgerby, photos by Lorenzo Vitturi **P. 94-95** ©Rene Gonzalez Architect, photos by Michael Stavaridis **P. 102** ©Fondazione Prada, photo by Attilio Maranzano **P. 103** ©Fondazione Prada, interior photo by Attilio Maranzano, exterior photo by Bas Princen **P.** ©Fraunhofer Institute for Wood Research, Wilhelm Klauditz-Institut (WKI), photo by Manuela Lingnau **P. 109** ©Sonogamma **P. 116-117** photos by Bartek Pekala **P. 118-119** ©Hotel V Nesplein **P. 121** ©Baswaphon **P. 127** ©Eric Staudenmaier **P. 128** ©Connie Zhou/OTTO **P. 130** ©Hunter Douglas Inc. **P. 131** ©Hunter Douglas Inc., photo by Ko Kuperus **P. 132-133** ©Creation Baumann **P. 135** ©Beyer Architekten, photo by Joern Lehmann **P. 136** ©Freudenberg **P. 138-139** ©Birdair, Inc. **P. 140** Sefar, photos by Eibe Sönnecken **P. 147** ©Fabric Systems **P. 148-149** ©Casalis **P. 152-153** ©Peter Johnston Architect PC, photos by Mark La Rosa **P. 154-155** ©Floto + Warner/OTTO **P. 156** ©Ecophon **P. 161** ©Robert Peck/Santa Fe Film and Media Studios, Inc. **P. 164-165** ©Smith & Fong **P. 173** ©Cannon Design, photo by James Steinkamp **P. 174-175** Photo by Naquib Hossain, available under Creative Commons ShareAlike 2.0/cropped + colour corrected **P. 176** ©SsD Images **P 178-179** ©Assemble, photos by Tanya Milbourne **P 180-181** ©Redshift Photography **P. 184-185** ©Ceilings Plus **P. 186** ©Epic Metals **P. 187** upper image ©Epic Metals; lower image ©VG Architects, photo by David Whittaker **P. 190-191** ©GEON Studio **P. 192** ©Formglas Products Ltd. **P. 193** ©Bernstein Associates; **P. 194-197** ©Topakustik/NH Akustik + Design **P. 198-199** ©Fantoni/FantoniUSA **P. 200-201** ©Smith & Fong **P. 202** ©dukta - flexible wood **P. 203** ©ZHdK, photos by Regula Bearth **P. 204-205** ©Organoid Technologies GmbH **P. 206-207** ©Flex Acoustics, photos by Niels W. Adelman-Larsen **P. 208** ©Sonogamma **P. 210-211** ©molo **P. 219** photo by Jean-Pierre Dalbera, available under Creative Commons Attribution 2.0 **P. 220** ©Newmat USA Ltd **P. 222-223** ©Newmat USA Ltd **P. 225** ©Nimbus Group **P. 230-231** ©Richter akustik & design, GmbH & Co **P. 234** ©Peter Field Peck **P. 240** Union Station waiting room photo by Duc Su **P. 241** Photo by Steve Hymon & Julie Sheer, available under Creative Commons Attribution 2.0 Generic **P. 242-243** ©BAUX **P. 244** ©Evergreene Architectural Arts **P. 245** ©Architectural Resources Group **P. 247** ©Evergreene Architectural Arts **P. 248-250** ©Anna Kyyro Quinn Studio **P. 253** ©Vescom **P. 254** ©Luperto/ Annette Douglas Textiles AG **P. 255** ©Annette Douglas Textiles AG **P. 256-259** ©Paul Tahon and R & E Bouroullec **P. 259** alcove sofa ©Paul Tahon and R & E Bouroullec; workbay ©Studio Bouroullec **P. 260-262** ©BAUX **P. 263** ©BAUX, photo by Johan Lindstrom **P. 264-267** ©Mark Cavagnero Associates Architects, photos by Jasper Sanidad **P. 268** photo by Anders Ingvartsen **P. 269** photo by Brady Peters **P. 270** photos by John Gollings **P. 271-272** ©Nigel Young/Foster + Partner **P. 280-281** ©Paul Tahon and R & E Bouroullec **Front cover image** Kvadrat Clouds 2008 by Ronan and Erwan Bouroullec ©Paul Tahon and R & E Bouroullec

FIGURES

P. 19 (ASTM E795) **P. 21** (LONG) **P. 32** middle figure: (DOELLE), bottom figure: (EVEREST) **P. 33** top figure: (EVEREST), bottom figure: (DOELLE) **P. 35** (DOELLE) **P. 39** top figure: (EVEREST) **P. 40** top figure: (EVEREST) bottom figure (HARRIS); **P. 41** (HARRIS) **P. 145** based upon manufacturer test data **P. 167** (EVEREST) **P. 168** based upon manufacturer test data **P. 169** (EVEREST) **171** (LONG) **P. 216-217** based upon manufacturer test data

REFERENCES / FURTHER READING

- T Cox and P D'Antonio, *Acoustic Absorbers and Diffusers: Theory, Design and Application* (Taylor & Francis; 2009)
- L Doelle, *Environmental Acoustics* (McGraw-Hill; 1972)
- D Egan, *Architectural Acoustics* (J. Ross Publishing; 2009)
- M Ermann, *Architectural Acoustics Illustrated* (John Wiley & Sons; 2015)
- F Everest and K Pohlmann, *Master Handbook of Acoustics* (McGraw-Hill; 2009)
- H Fuchs, *Applied Acoustics: Concepts, Absorbers, and Silencers for Acoustical Comfort and Noise Control* (Springer-Verlag; 2013)
- C Harris, *Handbook of Acoustical Measurements and Noise Control* (McGraw-Hill; 1991)
- T Jester, *Twentieth-Century Building Materials: History and Conservation* (The Getty Conservation Institute; 2014)
- V Knudsen, *Architectural Acoustics* (John Wiley & Sons; 1947)
- T Schultz, *Acoustical Uses for Perforated Metals: Principles and Applications* (Industrial Perforators Association, Inc.; 1986)
- E Thompson, *The Soundscape of Modernity* (The MIT Press; 2009)

ABOUT THE AUTHOR

Tyler Adams is an acoustical engineer based in Los Angeles, California. He holds a Master of Science in Architectural Acoustics from Rensselaer Polytechnic Institute and is a member of the Acoustical Society of America and the Institute of Noise Control Engineers. As an acoustician, he has consulted on a variety of projects including schools, hospitals, offices, high-rise buildings, research laboratories, performance spaces, and environmental noise.

Sound Materials
A Compendium of Sound Absorbing Materials for Architecture and Design

Publisher
Frame Publishers

Author
Tyler Adams

Production
Carmel McNamara

Art Direction and Design
Abraham Rivera

Prepress
Tyler Adams and Abraham Rivera

Trade distribution USA and Canada
Consortium Book Sales & Distribution, LLC.
34 Thirteenth Avenue NE, Suite 101,
Minneapolis, MN 55413-1007
United States
T +1 612 746 2600
T +1 800 283 3572 (orders)
F +1 612 746 2606

Trade distribution Benelux
Frame Publishers
Laan der Hesperiden 68
1076 DX Amsterdam
the Netherlands
distribution@frameweb.com
frameweb.com

Trade distribution rest of world
Thames & Hudson Ltd
181A High Holborn
London WC1V 7QX
United Kingdom
T +44 20 7845 5000
F +44 20 7845 5050

ISBN 978-94-92311-01-6

Printed on acid-free paper produced from chlorine-free pulp. TCF ∞
Printed in Slovenia.

987654321